5·19·11 56277 64.13
AM

Rethinking Family Practices

Palgrave Macmillan Studies in Family and Intimate Life

Titles include:

Graham Allan, Graham Crow, Sheila Hawker
STEPFAMILIES
A Sociological Review

Harriet Becher
FAMILY PRACTICES IN SOUTH ASIAN MUSLIM FAMILIES
Parenting in a Multi-Faith Britain

Elisa Rose Birch, Anh T. Le and Paul W. Miller
HOUSEHOLD DIVISIONS OF LABOUR
Teamwork, Gender and Time

Jacqui Gabb
RESEARCHING INTIMACY IN FAMILIES

Peter Jackson (*editor*)
CHANGING FAMILIES, CHANGING FOOD

Riitta Jallinoja and Eric Widmer (*editors*)
FAMILIES AND KINSHIP IN CONTEMPORARY EUROPE
Rules and Practices of Relatedness

David Morgan
RETHINKING FAMILY PRACTICES

Eriikka Oinonen
FAMILIES IN CONVERGING EUROPE
A Comparison of Forms, Structures and Ideals

Róisín Ryan-Flood
LESBIAN MOTHERHOOD
Gender, Families and Sexual Citizenship

Tam Sanger
TRANS PEOPLE'S PARTNERSHIPS
Towards an Ethics of Intimacy

Elizabeth B. Silva
TECHNOLOGY, CULTURE, FAMILY
Influences on Home Life

Palgrave Macmillan Studies in Family and Intimate Life
Series Standing Order ISBN 978–0–230–51748–6 hardback
978–0–230–24924–0 paperback
(*outside North America only*)

You can receive future titles in this series as they are published by placing a standing order. Please contact your bookseller or, in case of difficulty, write to us at the address below with your name and address, the title of the series and the ISBN quoted above.

Customer Services Department, Macmillan Distribution Ltd, Houndmills, Basingstoke, Hampshire RG21 6XS, England

---

Rethinking Family Practices

David H. J. Morgan
Emeritus Professor, University of Manchester, UK

First published 2011 by
PALGRAVE MACMILLAN

Palgrave Macmillan in the UK is an imprint of Macmillan Publishers Limited, registered in England, company number 785998, of Houndmills, Basingstoke, Hampshire RG21 6XS.

Palgrave Macmillan in the US is a division of St Martin's Press LLC, 175 Fifth Avenue, New York, NY 10010.

Palgrave Macmillan is the global academic imprint of the above companies and has companies and representatives throughout the world.

Palgrave® and Macmillan® are registered trademarks in the United States, the United Kingdom, Europe and other countries.

ISBN 978–0–230–52723–2   hardback

This book is printed on paper suitable for recycling and made from fully managed and sustained forest sources. Logging, pulping and manufacturing processes are expected to conform to the environmental regulations of the country of origin.

A catalogue record for this book is available from the British Library.

A catalog record for this book is available from the Library of Congress.

10   9   8   7   6   5   4   3   2   1
20   19   18   17   16   15   14   13   12   11

Printed and bound in The United States of America

Contents

Series Editors' Preface

The remit of the *Palgrave Macmillan Studies in Family and Intimate Life* series is to publish major texts, monographs and edited collections focusing broadly on the sociological exploration of intimate relationships and family organization. As editors we think such a series is timely. Expectations, commitments and practices have changed significantly in intimate relationship and family life in recent decades. This is very apparent in patterns of family formation and dissolution, demonstrated by trends in cohabitation, marriage and divorce. Changes in household living patterns over the last twenty years have also been marked, with more people living alone, adult children living longer in the parental home and more 'non-family' households being formed. Furthermore, there have been important shifts in the ways people construct intimate relationships. There are few comfortable certainties about the best ways of being a family man or woman, with once conventional gender roles no longer being widely accepted. The normative connection between sexual relationships and marriage or marriage-like relationships is also less powerful than it once was. Not only is greater sexual experimentation accepted, but it is now accepted at an earlier age. Moreover heterosexuality is no longer the only mode of sexual relationship given legitimacy. In Britain as elsewhere, gay male and lesbian partnerships are now socially and legally endorsed to a degree hardly imaginable in the mid-twentieth century. Increases in lone-parent families, the rapid growth of different types of stepfamily, the de-stigmatization of births outside marriage and the rise in couples 'living-apart-together' (LATs) all provide further examples of the ways that 'being a couple', 'being a parent' and 'being a family' have diversified in recent years.

The fact that change in family life and intimate relationships has been so pervasive has resulted in renewed research interest from sociologists and other scholars. Increasing amounts of public funding have been directed to family research in recent years, in terms of both individual projects and the creation of family research centres of different hues. This research activity has been accompanied by the publication of some very important and influential books exploring different aspects of shifting family experience, in Britain and elsewhere. The *Palgrave Macmillan Studies in Family and Intimate Life* series hopes to add to this list of influential research-based texts, thereby contributing to existing knowledge and informing current debates. Our main audience consists of academics and advanced students, though we intend that the books in the series will be accessible to a more general readership who wish to understand better the changing nature of contemporary family life and personal relationships.

We see the remit of the series as wide. The concept of 'family and intimate life' will be interpreted in a broad fashion. While the focus of the series will clearly be sociological, we take family and intimacy as being inclusive rather than exclusive. The series will cover a range of topics concerned with family practices and experiences, including, for example, partnership; marriage; parenting; domestic arrangements; kinship; demographic change; intergenerational ties; life course transitions; stepfamilies; gay and lesbian relationships; lone-parent households; and also non-familial intimate relationships such as friendships. We also wish to foster comparative research, as well as research on under-studied populations. The series will include different forms of book. Most will be theoretical or empirical monographs on particular substantive topics, though some may also have a strong methodological focus. In addition, we see edited collections as also falling within the series' remit, as well as translations of significant publications in other languages. Finally we intend that the series has an international appeal, in terms of both topics covered and authorship. Our goal is for the series to provide a forum for family sociologists conducting research in various societies, and not solely in Britain.

Graham Allan, Lynn Jamieson and David Morgan

Acknowledgements

As will become apparent this is a project which goes back to before 1996 and which has developed and evolved in one form or another since then. Inevitably, therefore, numerous people have contributed to my thinking on this subject in ways recognised and unrecognised. I should like to thank, firstly, Carol Smart and all her colleagues for being convinced of the interest of my work enough to want to name a centre after me. The thought of this action still touches me and I can only hope that this book justifies their recognition. Second, there are the numerous friends and colleagues in the British Sociology Association Family and Intimate Relations and Auto/biography Study Groups. These have been a continuing source of support and inspiration. Third, several colleagues at NTNU, Trondheim, provided me with a different national perspective on family studies. I am grateful, in particular, to Berit Brandth, An-Margritt Jensen, Elin Kvande, Kari Moxnes and Bente Rasmussen. A special 'thank you' to Graham Allan who read the manuscript as well as providing constant support along with Lynn Jamieson, the third editor of this series and Philippa Grand at Palgrave.

As always, Janet Finch has continued to provide support and lively conversation about family and many other topics. And many thanks to all the members of the families I am connected to in different ways. They have kept alive my belief that family life, however defined, continues to provide fascination and delight.

1
The Original Argument

Introduction

The objectives of this book are relatively straightforward:

- To explore the range of meanings and usages attached to the term 'family practices';
- To discuss the relationship between the term 'family practices' and the more general use of the word, 'practices';
- To consider possible ambiguities within and criticisms of this term and its usages and to explore how these might be met and developed in order to develop and enhance our understanding of family life and its place within wider social settings.

But why use the somewhat clumsy and unfamiliar term 'family practices' in the first place? Why not continue to refer, more simply, to 'the family' or 'family life'? In order to answer this preliminary question we need to go back to the point where I originally elaborated on the idea of 'family practices' in my book *Family Connections: An Introduction to Family Studies* (1996, Polity Press).

The aim of *Family Connections* was to develop a slightly new angle on sociological studies of family life. As someone who had been teaching family studies for several years I had become increasingly aware that family sociology was somewhat marginalised within the wider spectrum of sociological enquiry generally. Possibly this feeling represented a form of academic paranoia and perhaps most researchers feel that their chosen specialisms are being pushed to the margin by more fashionable topics. However, there was at the time of writing a diffuse sense that family studies had been superseded by women and gender studies and all that needed to be said about the family could be said within these developing, and certainly more popular, areas.

One response to this would be to argue that while gender is an important dimension within family life and while gender studies made a significant contribution to studies of relationships between family members, there was still more to be said that could not be wholly absorbed within these developing areas of enquiry. Although gender constituted an important lens through which to study family processes, it was not the only one. Topics that still demanded attention included kinship, inter-generational relationships and sibling relationships.

While recognising the importance of these, and other topics, my approach was somewhat different. If, instead of making the family and family relationships one's point of departure, one began with some other topic which might appear to have little to do with these domestic matters, one then might begin to ask novel or interesting questions which would highlight the abiding importance of family life. Thus, if one began with a major area of concern in British sociology, such as social class, one quickly came upon arguments to the effect that it was 'the family' and not the individual that was the key unit in systems of social stratification (Morgan, 1996: Chapter 2). Studies of work and employment raised questions about the relationships between attitudes and practices within the workplace and commitments and involvements outside. The 'life' in the phrase 'work/life balance' was normally understood to refer to family obligations. In a slightly different way, I argued that developing areas of interest to do with, for example, Time, Space and The Body could also be seen to have a family dimension. In short, in order to study family life you did not necessarily have to begin with those areas conventionally associated with the word 'family' such as parenting and partnerships.

The idea of 'family practices' appeared in the concluding chapter of *Family Connections* and it was only when I was thinking about what might appear in this chapter that I started to use this term. Having begun to use the term I wanted to try to explore what it really meant and how it might be used a little more systematically. In part, I wanted to develop a term that reflected the kind of fluidity that I was working towards in the earlier chapters when I was trying to show how 'family' was implicated in a whole range of other social institutions and sets of practices. But the term 'family practices' was also developed to address another set of questions.

These questions revolved around the very use of the term 'The Family'. There were frequent references to 'The Family' in titles of books (including two of my own), in sociology catalogues and journal articles as well as in political or religious pronouncements. Increasingly, from the 1960s onwards, the use of the term with the definite article in this way was seen as presenting a range of difficulties.

The first was the danger of reification or a kind of misplaced concreteness. To write of 'The Family' was to give it a thing-like quality. It failed to do justice to the range of positions (or 'roles') associated with the family and the different ways in which these might be interpreted or enacted. It failed to do justice to the ways in which families were constantly undergoing change, whether we are talking about families in general located in historical time or any individual family moving through that time. It failed to distinguish adequately between ideas about and understandings of the family on the one hand and actual day-to-day family living on the other. This contrast was memorably expressed in John Gillis's distinction between 'the families we live by' and 'the families we live with' (Gillis, 1996). To sum up a lot of complex arguments, there is no such thing as 'The Family'.

But this was not simply a question of making some intellectual mistake or of misdirecting scholarly enquiry. It also had political or moral implications. In public discourse, and perhaps in some scholarly discourse as well, there was the ever-present danger of giving the idea of 'The Family' some kind of normative status. Practices might be evaluated according to the extent to which they conformed to or departed from some standard model of the family. This was frequently referred to as the 'cornflakes packet' image of the family, the association between a particular representation of family life with everyday items of domestic consumption. This family consisted of a mother, a father and two children, one boy and one girl.

There is scarcely any need now to list all the things that are wrong with this particular image of family life. But there is one theme of considerable importance, which deserves more than passing reference, and that is the idea of heteronormativity. Notions of 'the family' had at their heart the idea of a heterosexual couple (married or cohabiting) and their children. The standard model linked, in a seemingly effortless manner, family, gender and sexuality. Not only were other heterosexual ways of living or doing family excluded or marginalised but the same was also true of gays and lesbians and their various living arrangements. While the standard model could accommodate deviations such as cohabitation and re-constituted families (separation and re-partnering) it seemed unable to go beyond these. The importance of this continues to have political and practical repercussions. There are dangers that the

Anglican Church will split over issues of gay marriage and gay clergy and these issues continue to emerge in American political life. In the 2008 elections, the State of California voted to rescind the previous recognition of gay marriages.

Thus the term 'The Family' not only oversimplified a large range of practices, statuses and experiences but it also carried some strong normative baggage that disadvantaged certain groups in society; not only gays and lesbians but also lone parents, couples without children and people living on their own for a variety of reasons. The question now was how to proceed?

One possible answer would be to abolish the word family altogether at least as far as it is possible. By all means talk about parents and children or different modes of partnership or intergenerational relationships. But, it might be argued, there is no need to bundle all these sets of relationships together under some over-arching label called 'the family'. In fact, very often when people use the word 'family' they are referring to relationships between parents and children so there would not be a great deal lost if the latter rather than the former term were more routinely used.

Such an approach would certainly be useful in gaining a greater degree of specificity and in avoiding the emotionally charged if analytically indistinct term 'the family'. However, something might be lost by such a strategy. There are occasions, as we shall see, when the more general term 'family' might be the most appropriate term even though the sets of individuals described might vary with each usage. A 'family event', for example, is distinct from some other event however similar in terms of activities. Further, it is clear that family life continues to be of importance to individuals, and family matters frequently enter into everyday conversations between friends or acquaintances. Again the events so described may be extremely various and involve different sets of people (a daughter's progress at university, a grandfather's recent fall, a forthcoming wedding) but they often merge into a more general discussion of family relationships. Further, the fact that there are affinities between these events and individuals reflect the fact that individuals frequently have to make choices, establish priorities, between different sets of claims (Finch and Mason, 1993). While individuals, apart from politicians or sociologists, may rarely talk about 'The Family', family is frequently woven into everyday conversations and routine concerns.

Another strategy, one suggested by Elin Kvande, in her discussion of gender, is to distinguish between 'family' as a noun and as an adjective, and as a verb (Kvande, 2007). We have already discussed the problems associated with talking about 'the family'. Pluralising the term to 'families' may take us some way and indeed this is probably the most common usage today. But there are still problems about the fuzziness of the borders between families and non-families and the possible heteronormativity built into this expanded usage. The next step is to treat the word 'family' as an adjective as in 'family life', 'family processes', 'family events' – and 'family practices'. Here we are using the term 'family' as a particular, but not exclusive, lens through which to describe and to explore a set of social activities. Implied here is a recognition that this is one lens among several. Or, to change the metaphor, 'family' is like a primary colour which is most useful when blended with other primary colours to produce something distinct from the constituent parts, 'family and class' or 'family and gender' for example.

In practice, much of this discussion in this and in subsequent chapters, takes this 'family as adjective' approach. But what about family as a verb? Strictly speaking, the word is impossible in the English language. 'I family' sounds like a poor translation. However, there is a strong stream in sociological writing which emphasises 'doing' rather than simply having or being. The title of Kvande's book refers to 'doing gender' (2007) and it is possible to similarly talk about 'doing family'. Even if it still sounds a little strange it reminds us that family is about process and doing and this, and more, is implied in the idea of family practices.

Practices

In this section I shall return to my original list of what I saw as the key implications of using the term 'practices', although not in exactly the same words (Morgan, 1996: 188-91). Obviously, I make no claim to have coined the term 'practices' (see the next chapter) and indeed the term 'family practices' was used on at least one occasion before my elaborations. However, what I think was, and remains, distinct is my attempt to spell out what is implied in the use of this term.

Linking the perspectives of the observer and the actor

Sociology has always been concerned with the difference between the perspectives and concepts developed by the researcher and the perspectives and understandings deployed by those who are under investigation. At first glance, use of the term 'practices' would seem to fall very much at the external researcher end of the continuum. After all, individuals do not routinely talk about engaging in family, or any other kind of, practices. They just do them and live them.

However, as my discussion of the term 'family' and its many usages suggest, the danger of ignoring the variety of ways in which family life is lived and experienced animated much of the concern about using the word 'family' as a noun. In what way is the use of the term 'family practices' an improvement? Simply using the term without elaboration or qualification almost certainly represents little real achievement. However, using the term in full recognition of the implications of its use, as outlined in what follows, may provide some genuine gains. In short, if the use of the term 'family practices' is of any value it is because it opens up the possibility of movement between the perspectives of the observer and the perspectives of family members. This should become more apparent when we consider the next two sections.

A sense of the active

I have already explored the possibilities of using the word family as a verb and while this might possibly be a step too far, we need to retain and elaborate a sense of family life as a set of activities. Individuals can, in other words, be seen as *doing* family. A whole set of what appears to be trivial or even meaningless activities is given meaning through its being grouped together under one single label, that of family. The focus on doing, on activities, moves us away from ideas of the family as relatively static structures or sets of positions or statuses. Family actors are not simply persons defined as mothers, fathers and so on but they can also be seen as 'doing' mothering or fathering.

A sense of the everyday

In using the term 'family practices' I intend to convey a sense of everyday life both in the sense of those life-events which are experienced by a significant proportion of any population (partnering, parenthood, sickness, bereavement) and, equally, those activities which seem unremarkable, hardly worth talking about. There are differences between these two meanings of the term 'everyday' (Morgan, 2004) but here I want to keep them together. The contrast here, in part, is with a more 'social problems' approach to family life, one which emphasises various forms of family breakdown or dysfunction. It also emphasises a sense of commonality of family experiences, one which in some measure cuts across differences in terms of class or ethnicity.

A sense of the regular

One meaning of the term 'practice' in everyday life is one which conveys a sense of the regular as when someone practises the violin or swimming. In family life these regular practices might not necessarily make perfect but they are certainly part of everyday family living. These regularities may be daily, weekly, monthly or annual. They may be shared by a large section of any particular population (the school-run, for example) or they may be peculiar to members of a particular household or set of family members ('family' jokes, rituals, etc).

A sense of fluidity

Here, to make a distinction which was probably not wholly clear in my earlier discussion, we are dealing with two linked senses of 'fluidity'. The first is in terms of the boundaries of any one set of family activities, as to who is included or excluded. Who 'counts' as family depends in part on who is asking the question (a researcher, a neighbour, a social worker) and on the circumstances in which the designation of family membership might be seen as important. (Feast days, weddings, notifications of death and so on.) Once again we are reminded that the term 'family' is a highly flexible one.

But I also intended to convey another sense of the fuzziness of the boundaries between family and non-family. Any sets of practices which we might like to describe as family practices might also be described in some other way, at least in part. Thus most, if not all, family practices might also be described in some measure as 'gendered practices' or 'generational practices'. Commuting might normally be thought of as a work or employment practice but, since the journey is between home and work, it may equally be seen as a form of family practice. Practices merge and overlap with each other like splodges of watercolour paint or those puzzles which can be seen at first one way and next some other way.

A linking of history and biography

To use Mills' (Mills, 1959) often-repeated formulation, this may possibly be rephrased as a linking of the public and the private although the two sets of terms are not exact equivalents. This is in part a recognition that individuals do not start from scratch as they going about family living. They come into (through marriage or parenthood, say) a set of practices that are already partially shaped by legal prescriptions, economic constraints and cultural definitions. This is, in part, what is meant by structuration as a set of processes rather than fixed external structures.

We may perhaps understand this strand further by considering another usage of the term 'practice' in the English language. This is when we talk about professionals and others as being given a 'licence to practice'. This identifies a particular set of exchanges between legal processes and the activities of individuals, one that defines limits but which does not necessarily prescribe in finest detail. Everett Hughes usefully developed the terms 'licence' and 'mandate' to analyse professional practices and I have argued elsewhere that these terms could be extended to family members (Hughes, 1971; Morgan, 2002). These six headings convey the key implications of my decision to write about family practices. Not much of this will come as a great surprise to most readers although I consider the detailed elaboration to be novel. Further, these different strands in my understanding of 'practices' should be taken together and as influencing each other. Thus, for example, the routine and the everyday may draw attention to the links between history and biography. A concentration on the 'everyday', however defined, may serve as a reminder that these elaborations are not simply the constructions of an external observer. If this elaboration of family practices is of any value it might be found in further enquiry into the interconnections between these strands.

I reserve more critical discussion to a later chapter. However, at this stage, it might be helpful to identify one possible omission in the original account and one further, but important, rider to this discussion. The possible omission is to do with the idea of reflexivity. Ideas of reflexivity themselves have different strands of meaning but here I am referring to two of these.

The first is to do with reflexivity on the part of the observer. Here I am concerned with the process by which the observer attaches the label 'family' to any set of practices. I shall explore this in more detail in the next section but it is useful to mention it here as it is so closely bound up with the first of these strands, the linking of the standpoints of the researcher and the researched. This is in part a recognition of the fact that the observer is not studying an area of life of which he or she has little experience (as might be the case in some forms of criminal activity, for example) but, in contrast, something in which the researcher is already deeply implicated. The danger is, frequently, that the observer 'knows' (i.e. has experienced) too much about family life before embarking on a particular piece of research. Reflection on the nature of practices and how they are defined may provide for a useful opportunity for the researcher to understand how he or she shapes what is being observed.

The other meaning of reflexivity is on the part of the observed. Like the researchers, the observed already have a great deal of everyday and practical knowledge of family living, or of any other designated sphere of living. Yet this is not wholly unexamined and family members frequently have the opportunity to and the materials with which to look reflexively upon what they are doing. Thus a parent may say 'I don't know if I doing the right thing' in relation to her teenaged daughter and family members may frequently operate with some notions of 'normal' family life. To some extent the material for these reflections may come from increasingly available public advice on family life or representations in the forms of soap operas or advertisements. But even without these, one might expect that some routine monitoring of whether one's practices come up to expectations as to what is proper or normal takes place.

The rider is an implication of these elaborations which I did not fully appreciate at the time of writing. This implication recognises that family practices need not necessarily take place in locations that have a strong identification with family. These are places described as 'homes' or 'households'. The idea of family is frequently given a particular location in social space and this may contribute to its reification. However, it takes little further reflection to realise that family practices, on the part of individuals or particular combinations of individuals may take place 'away from home'. Talk about other family members at the workplace may be a form of family practice just as family practices may take place on holiday locations or in department stores or supermarkets.

The 'family' in 'family practices'

While I have provided illustrations from family life in order to outline the argument, it is not wholly clear why what I have described are *family* practices as opposed to any other set of practices. Indeed, I have argued that one of the implications of this approach is a sense of fluidity that what might be described as family practices might, through a different lens, be seen as something else, gender practices for example. Any designation of a set of practices as family practices must be part of the movement in the negotiation, between the perspective of an observer and of the participants.

One possible point of departure is provided by David Cheal in his characterisation of family practices:

> Family practices consist of all the ordinary, everyday actions that people do, insofar as they are intended to have some effect on another family member.
>
> (Cheal, 2002: 12)

I am not sure whether this is wholly accurate. It might be possible, for example, to think of some family practices (presenting a unified front against some external agency, for example) which are not orientated towards other family members. However, the important point in this quotation is that it stresses some kind of relationality; family practices are carried out with reference to some other family member. But how do we define 'family members'? If we define a family member in terms of a member of some designated collectivity, we may find ourselves back at the point where we started, the difficulties of writing or talking about the family as a noun. However, if we define family members in terms of everyday practices which so designate them, the difficulty is of a different order even if it does not disappear entirely. There is a necessary circularity, or reflexivity, involved. The practices, including not merely what is done but also how it is done, define who counts as a family member, at least for the time that these practices are being followed.

This may seen unnecessarily complicated but let us follow the example used by Cheal, that of family talk (Cheal, 2002: 12). This is where family members exchange talk, exchanging knowledge and experiences through everyday conversations, including those carried out through the telephone, email or texting. For example, a sense of obligation to 'keep in touch' may encourage someone to telephone a relative. (This is not exclusively a family matter; friends may experience a similar sense of obligation.) In so doing, there is a sense of recognition of the other as a family member and this may continue as the conversation continues with reference to others who are equally understood to be family members. Thus family practices are not simply shaped by the sets of sentiments and obligations, accumulated over long periods of time, that define them to be about family and not some other set of relationships. A sense of family is itself reconstituted through engaging in these practices.

Whatever the motivation, whether through a sense of obligation or genuine liking or through more practical considerations, the practices directed to another family member also constitute that other as part of a broader family constellation. The performance of such everyday, often routine, family practices underlines a sense of distinctiveness involved in engaging in these particular activities. 'Distinctiveness' here does not necessarily imply some sense of priority or privilege although it can do so. It simply entails some kind of recognition that the set of people involved in any current set of practices are distinguishable as family and not as friends, colleagues, neighbours or whatever.

It should be clear that this somewhat laboured way of describing how family practices constitute family members as much as family membership directs family practices is rarely a matter of calculation. What I have done here is present family practices in slow-motion, a process which highlights and gives undue emphasis to what is normally taken-for-granted and unproblematic.

Family practices, therefore, normally can be seen as having this taken-for-granted quality. However, there are occasions when what is normally tacit may be made explicit and the constitutive character of family practices may be made plain. These are occasions when the routinised expectations and assumptions are breached or when, for some other reason, they become problematic.

A key example of this breaching is to be found in a couple of well-known experiments conducted by Garfinkel (1967). In the first experiment, students were asked to assume that they were boarders in their own homes and to describe these locations on the basis of these assumptions:

> Persons, relationships, and activities were described without respect for their history, for the place of the scene in a set of developing life circumstances, or for the scenes as texture of relevant events for the parties themselves.
>
> (Garfinkel, 1967: 45)

Many students found it difficult to maintain this particular stance. The next experiment went beyond description and asked the students to act out these assumptions for a short period:

> They were instructed to conduct themselves in a circumspect and polite fashion. They were to avoid getting too personal, to use formal address, to speak only when spoken to.
>
> (Ibid.: 47)

This might entail, for example, asking for permission to do things such as going to the refrigerator where such permission was normally taken for granted. While young people are frequently accused of treating their homes as hotels, these experiments showed the limitation of this popular understanding. Some students refused to act in this way and most of the others found the task extremely difficult. Family members were not amused when the nature of the experiment was explained to them.

Such deliberate breaching is one way in which the taken-for-granted flow of everyday family life may be revealed. Another set of occasions where the constitutive character of family practices might be, temporarily at least, called into practice is where there are transitions within family relationships over time. These transitions may be distributed along a continuum from the relatively scheduled to the relatively unscheduled. Examples of the former might include deciding when to treat the partner of one's son or daughter as 'one of the family'. Sometimes this might be stated explicitly; otherwise family practices directed towards this newcomer might provide for this inclusion. Less-scheduled transitions might include those following divorce and re-partnering, especially where there are other sets of hitherto unrelated children involved. Other breaches of routine family expectations might be where a son or daughter come out as gay or lesbian and introduce their partners and, possibly, their partner's family.

Thus, family practices are those practices which are, routinely or less-routinely, constituted as such. However, we also need to think of the processes by which the external observer constitutes a set of practices as being 'family' (and not some other) practices. In many cases, one may assume, there will be a considerable degree of congruence. This is where the observer is concerned with practices which are quite widely and conventionally associated with the idea of family: parenting and certain patterns of informal care across generations for example. Elsewhere, the 'family' character of practices may be less immediately obvious: changing house or job, buying a new car, joining a local campaign. Here, as in other areas of social enquiry, there will be some attempt to match, to establish congruence between, the perspectives of the observer and those of the participants.

Outline of the book

In this introductory chapter I have returned to my original presentation of the idea of family practices. To a large extent I have adhered to the original arguments and the order in which they were presented, although I have made one or two additions and provided some new illustrations. I have argued that any originality that might have been present in these arguments was in my attempting to spell out what was implied by using the term 'family practices'.

I am not sure whether there is any great originality in the term 'family practices' itself and certainly the term 'practices' has been around for some time. In Chapter 2, therefore, I attempt to locate the idea of 'family practices' within this wider discussion of practices, referring especially, of course, to the work of Pierre Bourdieu. Other usages of the term 'practices' will also be explored. In this chapter I shall also, briefly, refer to some other ideas that seemed to be dealing with somewhat similar problems to the ones that I have outlined earlier in this chapter. These will include the ideas of 'Caringscapes' and 'Family Configurations'. Since I presented my original argument in 1996, the term 'family practices' came to be used by several other researchers. I begin Chapter 3 by considering some of these usages as a way of fleshing out some of the more abstract ideas presented in this introduction. I also consider some extensions of the idea in recent writings of Carol Smart and Janet Finch. Generally, the references to and uses made of my work were gratifyingly positive. However, on reflection, I am aware of several possible gaps and ambiguities in the original argument. Thus, in the later part of Chapter 3, I engage in a form of auto-critique considering such issues as my possibly over voluntaristic bias and my underplaying of contextual factors. I also say a little more about gender (not ignored in my original argument but perhaps not as central as it might have been) and the relationship between practices and discourses.

The next three chapters take up some themes which have gained in attention within the wider field of sociology and which may be seen as especially helpful in developing the idea of family practices. These are the themes of 'Time and Space', 'The Body and Embodiment' and 'Emotions'. In each I consider how these provide different ways of enriching our understanding of family life as well as, reciprocally, how the family practices approach might contribute towards the understanding of these particular developing areas.

Chapter 8 also deals with some current developments in sociological enquiry but one which has a particular resonance in considering family life. This is what I call 'The Ethical Turn'. There have been some significant developments within family studies raising questions of ethics and encouraging a more distinctly sociological perspective on these concerns. Here I consider the extent to which the use of the ideas associated with family practices emerge out of these developments or might be considered to contribute to their further elaboration.

There is a slight shift in emphasis in Chapter 9. Issues of 'work/life balance' have been much discussed in recent years. In particular, there have been debates in several countries about such issues as parental/paternal leave and 'family-friendly' policies on the part of employers. There is a slightly more applied emphasis in this chapter where I consider the way in which the family practices approach might contribute to these policy debates.

In the concluding chapter I attempt to provide an overview of the preceding arguments and their various applications. I also attempt to take the argument a little further by raising methodological considerations. While the family practices approach is not committed to any one methodological orientation I shall argue here for a particular consideration of approaches using narrative and different forms of auto/biographical writing.

2
Locating Practices

Introduction

In the previous chapter I reiterated the various shades of meaning that I attached to my use of the word 'practices' and its particular application in 'family practices'. To a large extent, the first chapter confined itself to my earlier statements of these ideas.

In this present chapter I seek to locate the idea of 'practices' in general. This idea of 'locating' will be used in two senses. Here I shall attempt to locate the idea of 'practices in general' within the various theoretical traditions that have contributed to the elaboration of this idea and its application within a variety of spheres. In the next chapter I shall look more specifically at 'family practices' and my location will be attempted through reference to and comparison with some other recent approaches to the family and allied areas of social life.

This present chapter does not pretend to be a comprehensive theoretical overview of the developments within practice theory or of the use of the term 'practices'. Rather, the chapter is based upon a recognition that the idea of family practices did not develop in a vacuum; when I tried to set out what was implied by the use of the term, I was doing so in the context of numerous discussions, conference presentations and books and articles read, sometimes around similar and related themes and sometimes around themes which might have appeared at the time to be somewhat distant from my more immediate concerns. Some of these influences and discussions I have acknowledged; but many I have not and there are doubtless still other influences which I barely recognise and which exist at a more subliminal level.

It would be a lot neater if I were able to identify an author, or a body of texts, which influenced me when I first began to think about practices as applied to family studies. Certainly, there are several authors who have used terms such as 'practice', 'practices' and 'praxis' and some of these I shall be examining later in this chapter. But to say that there was a direct line from these authors or texts and my development of the idea of family practices would be, at best, misleading. In many cases, something of a reverse procedure might be identified. Having elaborated my ideas I then turn, or return, to these other authors and find some affinities between what they were saying and my own ideas.

What I am saying here, in effect, is that the term 'practices' and associated words were currently in use when I began to apply the word to family for the reasons I have explained in the first chapter. For example, I was aware that, in the field of cultural studies, people were talking about 'signifying practices'. This, I assume, entails a recognition that signs do not simply exist as things or social facts but that they are actively produced and modified by cultural producers and consumers. This diffuse sense of 'doing' rather than simply 'being' is something which I see as central to the idea of family practices. Thus the idea of 'practices' as a useful way of thinking about a variety of social processes was very much 'in the air' when I came to think about their application to family life.

'In the air' is unnecessarily vague and it is possible, at the outset and in a provisional fashion, to identify some more specific clusters of influences:

(a) Feminist thought and scholarship. This had a more specific association with my interest in family as it was part of a process that sought to remove any reificatory tendencies from the use of the term 'the family'. Thus critique was not simply applied to the family, however, but also to other widely used terms such as 'class', 'work', 'leisure' and so on, the aim in each case being to explore the gender differences and inequalities that were masked by the use of these terms and their deployment within sociological analysis.

(b) Ethnomethodology. Again, here was a critique of the general terms and categories deployed in sociological analysis (terms that often uncritically derived from more everyday usages) and which demanded that sociological inquiry should look at meanings in usage, the ways in which people 'did things with words' in everyday conversational and interactional contexts.

(c) Post-modern thought. Important here was not simply the critique of grand narratives but the development of a sense of fluidity in the alleged structures and processes which were supposedly the objects of sociological analysis.

(d) Various forms of reflexivity. This is another word that came to be very much in the air and overlapped with all of the other cited influences. Particularly important here is the recognition of the active role of the observer in the processes which were under investigation.

(e) The auto/biographical turn in social analysis. Various forms of auto/biographical writing and research based upon life stories and narratives have played an increasing part in recent social research. These approaches have, in turn, been influenced by the trends listed above. Such an approach is in opposition to an approach which sees social life in terms of variables and, in place, substitutes an approach which is both more holistic and more fluid and where time plays an integral part.

(f) One particular author, Pierre Bourdieu, is particularly identified with the idea of 'Practice' and 'practices' and I shall have more to say about him later. However, I do not, in the chapter, intend to go through particular authors and to provide a specific critical analysis of their different relevances. Rather I shall identify some key strands in the idea of 'practices' and see how they relate to these theoretical traditions and to particular authors located within or without these traditions.

Shades of meaning: Dictionary definitions

In this section I outline some of the key dictionary definitions which will constitute the basis for the analysis which follows. To a large extent these meanings derived from the dictionary correspond with many of the themes identified within the previous chapter. But some do not (and this may open up possibilities for further analysis) while one or two of my themes do not appear in the dictionary. All the quotations come from the most recent on-line version of the Oxford English Dictionary.

1. 'The carrying out or exercise of a profession'. Law and Medicine are seen to be particularly associated with this usage.
2. (a) 'The actual application or use of an idea, belief or method, as opposed to the theory or principles of it'.
2. (b) 'An action, a deed, an undertaking, a proceeding'. (Now merged with 3b.)
2. (c) 'The action of doing something: method of action or working' (Now merged with 3a).
2. (d) The Marxist or neo-Marxist approach as in 'Praxis'.

3. (a) 'The habitual doing or carrying on as usual, customary or constant action or performance, conduct'.
   (b) 'A habitual action or pattern of behaviour; an established procedure'.
   (c) 'An established legal procedure'.
4. 'Repeated exercise in or performance of an activity so as to acquire, improve or maintain proficiency in it'.
5. Here we find a cluster of obsolete or rare usages to do with scheming, conspiracy, collusions or stratagems. Whether these usages have much to do with family practices is a matter for some further discussion.
6. A practical treatise, an exercise for students (Rare).
7. A specialised mathematical usage.
8. 'The action of deluding, manipulating or deceiving someone' (Obsolete).

It can be seen that most of these usages overlap with those outlined in the previous chapter (especially 2 and 3) while others seem to be of limited relevance (3(c), 6 and 7). Others may be found to have some relevance and, if so, that is the value of this exercise. At the same time there were certain themes that were raised earlier which are not found here. This includes the idea of fluidity and the contrasts between the perspectives of the actors and the observers. This is because these are not definitions as such, but elaborations of the actual usages or implications of the term 'practices' as it has developed in sociological thought.

The exercise of a profession?

At a first glance this definition of practice, referring in particular to the legal and medical professions, would not seem to have much relevance for the discussion of family practices. Indeed, the worlds of occupations and professions, based on models of rationality, would seem to be at some distance from everyday understandings of family life. More generally, it would seem that the whole field of what has come to be called 'practice theory' is not primarily concerned with more specialised notions of professional practice.

This may possibly be a mistake. Returning to family practices there are various ways in which professional life and practice impacts upon them. In the first place, a variety of professions and professional orientations have developed which either focus primarily on the family as a focus of practice or which seek to develop a family dimension to a wider professional orientation. One thinks of Family Law, Family Courts, Family Therapy, family-based medicine and so on (Clark & Morgan, 1992; Morgan, 1985). These particular spheres of interest have grown over a considerable period of time and are now seen as relatively commonplace.

At the same time it can be argued that this professionalisation of family life is not confined to external interventions, based upon constructions of normal or abnormal family practices, but also enters the everyday practices carried out by family members. Thus, through attending classes, consulting advice books or reading articles in newspapers, individual family members monitor their everyday practices in terms of notions as to what it is to be a proper mother or father, wife or husband.

Some understanding of family practices may be gained by considering an influential sociologist in the field of the study of occupations and professionalisation, namely Everett Hughes (Hughes, 1971; Morgan, 2002). Several of Hughes' ideas might prove to be of use in analysing families (the idea of 'dirty work', for example) even if the word 'family' hardly occurs in his collection of essays on occupations and professions. But it is his discussion of 'licence' and 'mandate' which may be especially useful in thinking about family practices. Without going into great detail (see Morgan 2002), both terms refer to the kinds of exchanges that take place between family life and wider state and professional institutions. We are talking about the kinds of expectations that are placed upon family members in terms of their responsibilities for others and the kinds of claims which are made by family members for some kind of specialised understanding and set of rights that derive from their family positions. We can see these understandings being negotiated in, for example, long-running debates about education of children and the competing or overlapping claims of families and schools.

I hope that I have said enough to indicate that this first definition of practice does have some relevance for the discussion of family practices. Very generally it refers to the exchanges over time that take place between family members and groupings and state and professional institutions. Hence it relates to the links between biography and history referred to in the previous chapter.

Theory and practice

The second cluster of definitions focus upon practice as a form of action, the key words being 'actions', 'deeds', 'undertakings', 'proceedings', 'doing' or 'working'. However some of these more specific definitions are merged with definitions that occur in the next cluster dealing more with habitual or customary action and hence these will be considered in the next section. The opening definition, and presumably the one seen as being of prior significance, deals with the contrast between theory and practice. This is a contrast deeply enshrined within, one might suggest, British culture with its supposed suspicion of abstraction and theory and its endorsement of the practical and of everyday experience. 'That's all very well in theory ...' is a common enough opening phrase.

However, it would be wrong to confine this contrast to the familiar one between more or less abstract, specialised and systematised theory and the more everyday, complex and messy engagements in everyday life. The contrast is more complex than this. Consider, for example, Garfinkel's use of the word 'practical' which occurs three times in the opening sentence of his *Studies in Ethnomethodology* (1967: 1), where the reference is constantly to 'practical action' and 'practical reasoning'. Consider also Giddens' reference to 'practical consciousness' where he relates this to agents' 'knowledgeability as agents' (1984: xxiii). Certainly the contrast is in part between the constructions of sociologists and other theorists who provide models of everyday life and the lives and accounts of agents going about their daily life. But the contrast is also between the everyday forms of public accounting that social actors routinely engage in during their encounters with others and the ongoing everyday flow of life in families, organisations, leisure activities or whatever. Garfinkel's quest here is not one of exposing what lies beneath the surface but rather in highlighting a necessary contrast between what we say we are about and what we are actually doing. This is shown in the experiments conducted by him which showed the impossibility of providing a full description of what was 'actually' going on. Put another way, theorising is necessarily implicated in practice. This inter-dependence between accounts and practices is stated by Schatzki when he writes that '... practices are the site where understanding is structured and intelligibility ... articulated' (Schatzki, 1996: 12).

This complexity is further illustration when we consider Reckwitz's contrast between 'practice' (or 'praxis') and 'practices' ('praktiken') (Reckwitz, 2002: 249). The former refers to the 'whole of human action' (which itself is contrasted with 'theory') while the latter, practices, refers to 'a routinised type of behaviour which consists of several elements, inter-connected to one another'. Here, 'practice' is itself a form of abstraction in contrast to the more routine practices. What these discussions point to is a more complex notion of the 'theory' which is contrasted with 'practice', one which does not privilege one over the other but which shows their necessary interconnection. In a slightly different way, this contrast between 'practice' and 'practices' highlights a certain ambiguity when 'practice' is contrasted with 'theory'. In one sense, when we are talking about 'practices', we are looking backwards in time to the extent that the practices exist prior to theory and the theory seeks to explain, understand and describe already existing practices. (Over time, of course, these 'already existing' practices will themselves be influenced by theory.) When we are talking about 'practice' or Praxis, we are talking about the way in which theory and theorising informs practical action (individual or collective) in the world in order to bring about some change in this world. Here, in a sense, we are looking forward from the theory to the practice.

The most celebrated example of this, of course, is to be found within Marxism and the contrast between explaining or describing the world and changing it. In terms of family studies, although by no means confined to these studies, the key example is provided by feminist writings. Looking backwards, to the actual lived experiences of women in family relationships and households, existing theories were found to be inadequate. Looking forward, the call was for ways of engaging in the world and using these new understandings of the world in order to bring about changes in structured gender relationships within the family and elsewhere (see, for one of many examples, Smith, 1987).

The distinction between theory and practice is apparent in the work of Bourdieu and his writings on 'Practice'. Bourdieu was clearly influential in my deployment of the term 'family practices' although the connections in my discussion tended to be implicit rather than formally worked out. When he talks about 'practices' he is, in some measure, talking about what is practical. We see this in the contrast between official and practical kinship. On the one hand there is an understanding of kinship in terms of a 'language of prescriptions and rules' (Bourdieu, 1990: p. 163). This understanding, derived from the more formal models drawn up by anthropologists or other external agencies, is seen as inadequate. We need to move closer to an understanding of 'practical kinship' and 'practical relationships' (Bourdieu, 1990: p. 168). A statement of the rules governing, say, the choice of marriage partners, will be an abstraction, one which is contrasted with the everyday practices shaped by practical considerations and immediate concerns. Another way of thinking about this contrast is in terms of the one contrasting 'rules' (externally constructed, abstracted models) and 'strategies'. This is, for example, one that Finch and Mason deploy in their analysis of family obligations, a study which has very much influenced my approach to family practices more generally (Finch & Mason, 1993). Here the emphasis is upon the idea of 'negotiation' a term which has some affinities and overlaps with the idea of 'strategies'. This approach is not, it should be stressed, one which rejects theory and abstraction. Bourdieu's whole project here is one which seeks to overcome the distinction between objectivism and subjectivism. Clearly the accounts that Bourdieu provides of his fieldwork are not identical with the accounts provided by people being investigated. Or rather, social actors do not routinely provide accounts until required to do so by outsiders; they simply live and do families and kinship. The use of terms like 'practices' and 'strategies' is itself one that deals in abstractions which are rarely routinely deployed in everyday life. What might be argued is that these abstractions, necessary in any form of social enquiry, are of a different order from and raise different questions to those abstractions previously deployed. Thus the search for the rules of kinship in a particular context may obscure the fact that in everyday practical life issues of kinship, family relatedness, are woven into other practical considerations to do with earning a living, with structured inequalities in terms of class and gender and, more generally, with making use of the limited resources (economic, social and cultural) that are immediately available. It is not so much, therefore, a question of replacing the abstractions of social scientific language with some kind of unmediated access to other peoples' lives (if, indeed, that were possible) but more one of replacing certain kinds of abstractions with others that are more capable of dealing with the complexities of everyday life.

Action

The definitions in 2(b) and 2(c) above refer to 'an action, a deed, an undertaking, a proceeding' and 'the action of doing something, method of action or working'. The dictionary definitions present these as being 'merged' with 3(b) and 3(a) but, for the present purposes it seems desirable to keep these separate. The latter sets of definitions deal with questions of 'habit' or established procedures and different issues seem to come to the fore here.

The term 'action', especially when it is understood as 'social action' is one which is at the heart of much sociological enquiry. Social action is defined, classically as human action which is in some way directed to the expectations of, or in recognition of, others. These 'others' may not necessarily be co-present individuals but may be anticipated (as in preparing yourself for an interview) or exist in the past (one's forebears or ancestors). They may be collectivities, real or imagined (reference groups, generalised others) or identifiable individuals. It can be readily seen that this understanding of action as social action is very much in keeping with the idea of family practices. Cheal, it will be remembered, defined family practices as 'all the everyday actions that people do, insofar as they are intended to have some effect on another family member' (Cheal, 2002: p. 12).

This understanding of practices in terms of action, social action, has affinities with the discussion about the relationship between theory and practice. The implicit contrast would seem to be between action and doing on the one hand and descriptions of and discourses about practices and the institutions within which they take place, on the other. The implications of this distinction between practices and discourses will be explored in a later chapter but it should be clear that the two are mutually implicated in each other. Discourses are a form of practice just as practices are given meaning and shape through discourses. Nevertheless it is a useful distinction to make at the outset especially as there is a tendency, in the case of family studies at least, for the discourses to be mistaken for the practices. Thus accounts of the family in modern society will begin, and sometimes end, with statistics about divorce, lone-parents, childless couples and so on. The range of practices indicated by these statistics is sometimes forgotten.

In a variety of metaphors, Bourdieu not only returns to this distinction between theory and practice discussed in the previous section but also shows how practices, everyday pieces of action, are necessary for the maintenance of relationships and social institutions. Thus, he writes:

> Official relationships which do not receive continuous maintenance tend to become what they are for the genealogist: theoretical relationships, like abandoned roads on an old map.
>
> (Bourdieu, 1977: p. 38)

Elsewhere, he deploys the same metaphor while reminding the reader of the difficulty of renewing a relationship that has not been maintained by letters or gifts (Bourdieu, 2008: p. 137). Yet again, he is fond of likening family practices, such as 'the matrimonial game', to a card game. We can only get a little way by consulting the book of rules; what really matters are the hands that have been dealt and the skill of the players (Bourdieu, 1977: p. 58). These metaphors do not simply restate the point about the contrast between theory and practice; they also highlight the central importance of practices for the continual maintenance and reproduction of the institutions under consideration. Popular statements about the necessity of 'working' at one's relationships also reflect this deeper understanding.

A focus on action and doing also, inevitably, entails a focus on the social actor and the idea of agency. We are clearly moving into some core debates within sociology, the relationship between agency and structure, and it is doubtful whether any brief discussion here could contribute much to these wider accounts. A focus on practices, in the sense of action and doing, would also seem to entail a focus on the actor rather than on structures or institutions. This may be a matter of moral principle, of personal preference or of methodological expediency or possibly of all three. I shall turn to the question of whether the practices approach entails a neglect of or a playing down of structure in a later chapter. Here I should note that to talk of *social* action and *social* actors (rather than using the more familiar term, 'individuals') already flags up some rudimentary (at least) sense of structure and structuration. Practices are essentially and inescapably relational even when we refer to, in Schatzki's phrase, 'the hanging-together of human lives' (Schatzki, 1996: p. 171). Action is conducted in relation to 'others' and these actors cluster together in some mutually understandable set of relationships to which the word 'family' (or whatever else we are interested in) might be attached.

A focus on action, at least within the Weberian tradition, has also frequently entailed attempts to distinguish between types of action. Within Sociology it would be fair to say that practitioners have seemed more at ease with some version of rational action even where this does not necessarily entail a signing up to a strong version of rational choice theory. Within family studies, for example, even where there might not be any direct reference to this body of theory, the deployment of terms such as 'strategies' or 'negotiation' might entail more than a nod in the direction of rational action, even if the rationality is more in terms of the analyst's attempts at understanding family life. Purely affectual or emotional action has tended, until fairly recently, to be placed outside the ambit of *social* enquiry. Habitual or traditional action, on the other hand, has received some attention and this brings us to our next cluster of definitions.

Habit

The third cluster of definitions refer to habit, 'the habitual doing or carrying on of something usual', 'habitual action or pattern of behaviour', 'established procedures'. The connotations are to do with what is seen as normal or routine, what is repeated, what is taken for granted. In many ways this would seem to be especially relevant when considering family practices with its daily, weekly, monthly or annual repetitions and cycles.

Habits may, very simplistically, be described as being either personal or collective, a distinction that probably has deeper meaning in modern and late-modern societies than in earlier social formations. Personal habits are attached to individuals and are often seen as marks of individuality, whether they are merely eccentric or rather unpleasant. Thus an individual might routinely remove his shoes on entering the house while another might equally routinely wander in without even removing any mud that may be attached. Some habits may come close to forms of obsessive behaviour (double checking that the door has been locked) while others might be closer to knowing, comic performances. As a student I once lived in a house where the father would frequently look in the local evening paper and announce the death of some individual. 'Who's he?', someone at the supper table would ask and he would reply, 'I don't know.'

Sociological discussions of habit, or Bourdieu's 'habitus', tend to be less concerned with these more idiosyncratic forms of behaviour and more with collective habits. Yet these personal forms have their place in the analysis of family practices as they enter into the ways in which family members understand themselves and their connections with each other. The habits may be personal but the responses to them and the meanings assigned to them are more collective.

Fowler describes Bourdieu's view of action as a 'subtle blend of experience and cultural unconscious' (Fowler, 2001: p. 321). One might add, especially in relation to habitus that this is more than a blend but an ongoing interaction between the two. One of Bourdieu's own memorable phrases in relation to habitus is that of practices which are 'collectively orchestrated without being of the orchestrated action of a conductor' (Bourdieu, 1977: p. 72). If we are talking of family practices we may perhaps be referring more to a small jazz or chamber music group rather than a full-scale orchestra. The point is that the action remains at the level of the unspoken or taken for granted: '... the system of schemes orientating every decision without ever becoming completely and systematically explicit ...'(Bourdieu: 2008: p. 135).

This taken-for-granted character is illustrated in another striking phrase when he talks about 'genesis amnesia' (Bourdieu, 1977: p. 79). Any set of practices, however habitual they may seem, have their historical origins; they do not simply 'just happen' and they are not simply the product of any one individual. The very idea of 'habitus' seems to entail the forgetting or, sometimes, the misrecognition, of the origins of the practices concerned. There is, therefore, an on-going, even circular, character about the practices which are included with the idea of habitus as these quotations indicate

> the system of dispositions inculcated by the material conditions of existence and by family upbringing (the *habitus*), which constitutes the generating and unifying principle of practices, is the product of the structures which the practices tend to reproduce.
>
> (Bourdieu, 2008: p. 134)

> Ordinary marriages, arranged between families united by frequent, long-standing exchanges, are marriages of which there is nothing to be said, as with everything that can be taken for granted because it has always been as it is. They have no other function, apart from biological reproduction, than the reproduction of the social relationships that make them possible.
>
> (Bourdieu, 1990: p. 181)

Habitus, therefore, is linked to his approach to social and cultural reproduction: 'each agent, wittingly or unwittingly, willy nilly, is a producer and reproducer of objective meaning' (Bourdieu, 1977: p. 79).

I have cited these quotations at length because they can be seen to highlight a tension within the different definitions of 'practices', the one focussing on action and the other focussing upon habit. If the emphasis on doing seems to highlight the activities of the social actor, and hence, agency, the focus on habit seems to lead us back strongly, even deterministically, towards social structure. The tension between these two sets of meaning of 'practices' is possibly a fruitful tension and one which will be explored later.

In exploring this tension it is important not to lose sight of the social nature of habitual action. Even though it may appear to be otherwise, it is not simply a form of conditioned reflex. Giddens refers to the process of 'routinisation' (Giddens, 1984) and sees routine 'as the predominant form of day-to-day social activity. Most daily practices are not directly motivated' (ibid.: 282). But this is not to say that they are not enacted by knowledgeable agents. When called upon to do so, by circumstances or through outside intervention, these agents are frequently able to account for these practices and to recognise their taken-for-granted quality.

Other meanings

Of the other meanings listed in the dictionary, perhaps the only one deserving further examination is the one which refers to 'repeated exercise of a performance of an activity so as to acquire, improve or maintain proficiency in it'. From the perspective of family practices, this definition has two elements. The first refers to the repeated character of the practices while the other refers to the aims of or consequences of carrying out these practices. Practice makes perfect.

The repeated character of family practices has already been mentioned under the section 'habit', and repetition was referred to in my original discussion of family practices. It is perhaps worth pointing to a difference between repetition and habit here. 'Habit' does not necessarily refer to repeated practices, although it may well do so. It refers more to the framework of reference in which certain practices are conducted. These practices may seem familiar because they come from a similar source but they are not necessarily repeated, at least not in every detail. Thus shopping may be a repeated act but, over the years, its character might change considerably. A daily trip to the shops may be replaced by a weekly visit to the supermarket. This may in turn be replaced by a regular order over the Internet, supplemented by irregular visits to a local store. What is common to all of this is not simply the carrying out of transactions, consumption practices, but the abiding sense of, say, shopping not simply for oneself but for a shifting set of others.

We might also note another difference between repetition and habit. Habit is something that is built up over many years and, possibly, over several generations. Moreover, it can refer not simply to a particular family or household but to a particular community or identifiable group such as a religious or ethnic group within a particular setting. Such habit, as the quotations from Bourdieu in the previous section indicate, is something that 'comes naturally', that is done 'as a matter of course', as representing how we usually do things. Repetition may have this character but it may also have a more overt, rational or strategic character as something that is clearly orientated to a particular end. Regular exercise, dieting, taking vitamins and so on may all have this character of repetitions deliberately carried out with certain ends in mind.

How far is this more 'rational' understanding of repeated practices relevant to our discussion of family life? We can think of two possible sets of relevance. In the first place it might be argued that much of this repeated activity is either carried out with reference to or on behalf of other family members (certain food practices or health-related activities, for example) or may have consequences for other family members. A regular jog, for example (perhaps in preparation for a local marathon or a charity run), has implications for the timetables of others to whom the jogger might be related. In the second place, an increasing focus on these kinds of activities – repeated and goal orientated – might be seen as part of a process whereby the home is subjected to increasingly rationalising tendencies (Reiger, 1985). This in part is a reflection of the involvement of external agencies (the state, professions) in everyday family practices. Thus, for example, parenting comes increasingly to be seen as a set of skills, which improve with practice, or at least practice that is informed by knowledge. One of the arguments in favour of paternal leave, for example, is to encourage fathers to develop skills of childcare which may be deployed over the life-course.

In fact, it might be supposed, most of the repetitions of family life consist of a mixture of the more habitual or traditional and the more rational or purposive. Regular holidays, for example, might be part of the normal expectations of family living, reflecting knowledge of what is done in other families and memories from previous generations. But they may also, perhaps encouraged by articles in magazines or newspapers, be seen as events that should not be left to chance and which should be seen as providing opportunities for 'doing' or displaying family.

There are, perhaps, two further definitions that might deserve some passing attention. The first is 3(c), 'an established legal procedure'. At the very least this may serve as a reminder of the various ways in which law and family are intertwined with the law, to some extent, reflecting widespread assumptions about family roles and responsibilities while, at the same time, also shaping and influencing these everyday domestic practices. We are reminded that family practices are not simply the outcome of individually negotiated decisions or strategies carried out in isolation from other areas of social life. We shall be returning to this in a later chapter.

Another area which may not seem to have a great deal to do with family practices are the obsolete or rare definitions cited in 5 and 8 which refer to scheming, conspiracy, collusions or stratagems. But a moment's thought might suggest that these definitions may have everything to do with family life. Seen from within, the politics of everyday family life might be seen or experienced as a sequence of schemes, stratagems and alliances.

Exercise of everyday power (over spouses, over children, over siblings) may routinely require all of this as individual's attempt to get other family members to do what they might not wish to do. In some cases these everyday family politics may have dark or unfortunate consequences or may be adjuncts to systematic patterns of abuse. More benevolently, they may simply be seen as means of getting by as a parent, a partner or a child. It would be difficult to imagine families without secrets and not all of these secrets will be the basis of explosive or tragic outcomes.

Similar considerations may arise when looking at particular families from the outside. Particular families, or families in general, may be perceived as engaging in collusive practices against external bodies such as the education system, the police, social workers and so on. Notions of 'the family comes first' or practices of 'amoral familism' may be seen in opposition to other systems or other sets of values, as obstacles to individual or social change. In many cases issues of social class or other divisions may enter into these practices involving families in seeming opposition to external agencies.

At the very least these definitions serve as a reminder of the more dysfunctional (as seen from one or more perspectives) aspects of family practices. They may be obsolete or rare in terms of linguistic usages but from the family practices perspective they may provide another, and important, entry into the everyday working of these practices.

Concluding remarks

In this chapter I have attempted to locate my elaboration of the idea of 'family practices' within various sets of usages of the term 'practices' and to explore how these different usages have contributed to my understanding. These 'usages' have been both the range of usages identified in the dictionary (which themselves reflect a range of usages both everyday and more specialised) and the way in which the word 'practices' has been used within more theoretical sociological literature. I began by reflecting that, when I started thinking about 'family practices', the term 'practices' was very much in the air and used quite frequently within a wide range of more or less specialised areas. My original deployment of the concept, recognised that I was using a term that was already being widely used within the social sciences (although rarely in relation to family life) and that I was simply attempting to explore these meanings as they applied to doing family.

This exploration, reflecting on both the dictionary definitions and the theoretical traditions that have informed the idea of practices, has produced some provisional findings. The implications of some of these will be elaborated in later chapters.

The first and most obvious finding is the range of definitions of the word practice (and associated words such as practices and praxis) to be found in the dictionary. These strands identified in the dictionary may also be found, in different ways and with different emphases within a range of theoretical approaches. Furthermore, this range is not greatly at variance with the components of family practices which I originally elaborated. At the same time this exploration did not simply mirror my original analysis. One of the virtues of this kind of exploration is that it highlights certain tensions and ambiguities within the cluster of definitions and approaches associated with practices. One of these is between the idea of practices as action, stressing doing and, in some measure, agency, and the idea of practices as habit. Some notions of habit and habitus seem to imply an unreflexive participation in traditions and practices which are not obviously owned by the actors and where any sense of agency seems, at best, muted. I have argued that this perception may be misleading. However, this is a tension which I did not identify in my original discussion.

Another tension or ambiguity occurs in the idea of repeated activities. In one sense the repeated activities seem to be very much identified with the ideas of habit and hence conform to models of traditional or even affectual action. Yet repetitions can be undertaken knowingly and rationally, orientated towards some identifiable goal such as health and general well-being. I have already noted that various approaches to family life often use metaphors which imply some measure of rationality in this sense; discussions in terms of strategies and negotiations, for example. The relationship between rationality and emotions will be developed in Chapter 7.

Some of the definitions of practices (those referring to law or the professions, for example) may draw attention to the ways in which family practices do not exist in isolation from other areas of social life but interact with and are shaped by these other areas. In part this is to do with the relationships between practices and discourses, the discourses in this case being those elaborated with professional or legal frameworks. This was something that was underplayed in my original account and will receive further attention in this present volume.

Finally, the 'obsolete' or 'rare' definitions of practices involving stratagems and deceptions which occur at the end of the dictionary account serve as a reminder of some of the less welcome or 'dysfunctional' aspects of family life. In my original discussion there was some danger of ignoring what I stressed in an earlier book on family theory (Morgan, 1975), namely the 'darker' and unheralded aspects of family life. Here I am referring not simply to the more obvious manifestations of these downsides of family living, child or partner abuse, psychological damage and so on, but also to the more double-edged politics of everyday family living. Again, I hope to have the opportunity to explore these later in this volume.

Hence there are clear affinities between wider theoretical discussions of practices, elaborated dictionary definitions and my elaborations of family practices. As I have argued, this should not come as a surprise since the 'practices approach' was very much present when I explored what this might mean in terms of family living. However, it is worth identifying some areas in my elaboration which have not been emphasised so far in this chapter. The first is, when I wrote:

> I seek to combine, in ways which may not necessarily be harmonious or completely congruent, the perspectives of the actor and the observer.
>
> (Morgan, 1996: pp. 188–9)

In different ways the strands discussed here, Bourdieu, ethnomethodology and feminist writings, all recognise and address this tension. The important thing, at the outset, is to recognise that there is a tension between these perspectives and that the solution is not to opt for one or the other. While a 'family practices' approach might well be closer to the ways in which family life is actually lived and experienced, it is not faithful repetition or recording of these experiences. Such an approach, as the Garfinkel experiments remind us, would be impossible anyway, and does not represent a reasonable aim for family studies.

My discussion of the 'fluidity' of the term 'family practices' (Morgan, 1996: p. 190) does not emerge fully in this present chapter. But this sense of fluidity should be implied by the use of the numerous dictionary definitions which, of course, are not specifically anchored to 'the family' at all. Further, an ethnomethodological approach would presumably seek to avoid any strictly prescriptive definition or bounding of the family and would treat constructions of the family as a topic for further investigation rather than a simple basis for family analysis. Bourdieu's approach (although he is often critical of what he views as ethnomethodological perspectives) would seem to imply some kind of fluidity in that his investigations are conducted with a recognition of the fact that kinship strategies, for example, are located within and given meaning by wider patterns of inequality and power differentials. Yet again, the feminist approaches would be sharply aware of the overlap between family practices and gender practices. My sense of the fluidity of family practices serves as a reminder of the dangers of more reified constructions of 'the family' while drawing attention to interesting issues that arise when, say, family practices overlap with consumption practices or class practices.

Finally, in my original discussion I draw attention to the way in which family practices (or practices in general) constitute links between history and biography (Mills, 1959; Morgan, 1996: p. 190). On reflection, I am not sure whether this is an inevitable outcome of a 'practices' approach and, indeed, there is a danger that history may take a back seat in the analysis. However, the approaches of Bourdieu and feminists, at least, would be orientated by such an understanding and insofar there was some imbalance in my original account this is one that I hope to correct here.

There is one remaining issue that I explored in my original account but which is only touched upon to date in this chapter. Up to now I have tended to refer (following the dictionary definitions) to 'practices in general' rather than to 'family practices" and where family matters have been referred to it is largely as illustrations to the more general discussion. In my original discussion, however, I did devote a few pages to discussing what is 'family' about 'family practices' (Morgan, 1996: pp. 191–7). I shall be returning to this theme at various points in the following chapters as well as in my concluding chapter. Here, I should simply point to the way in which family practices orientate themselves to 'others' who are defined in some way as family 'members'. (This is not strictly true as the discussion of 'family display' in Chapter 4 makes clear.) Thus family practices are not simply practices that are done by family members in relation to other family members but they are also constitutive of that family 'membership' at the same time. This relational approach serves as a valuable corrective to what might otherwise be a more individualistic emphasis in discussions of 'practices in general'. I would argue, therefore, that a discussion of family practices is not simply an application or illustration of 'practice theory in general' but an approach which can in some ways illuminate these wider discussions. I hope that this will be demonstrated in subsequent chapters.

3
Locating Practices – Alternatives

Introduction

In the previous chapter I attempted to locate the idea of family practices in terms of some of the different theoretical approaches that have informed this approach as well as in terms of the range of definitions of the term, 'practices', to be found in a dictionary. The emphasis was largely on the idea of practices in general, with family serving largely as a source of illustrations for these more general ideas. Here the emphasis will be more specifically focussed on 'family'. Here I shall look at some recent, or fairly recent, approaches which might be seen as alternatives or complementary to the idea of family practices.

In the early statement of the idea of family practices (e.g. Morgan, 1996), some attempt was made, directly or indirectly, to locate this approach in relation to some alternative approaches. These alternative approaches, while differing between themselves, frequently had one feature in common. This was to treat the family as a unit or possibly a structure which could readily be expressed as a noun, *the* family. On this basis analysis could proceed to ask questions about the family in modern society, change in or the decline of the family or about the relationships between the family and other structures or areas of social life such as the economy, the class structure and so on. Putting to one side some phenomenological perspectives (which influenced then and continue to influence my thinking on this topic), the underlying assumption was that it was possible to talk about the family as a relatively clearly bounded unit and that further analysis could continue on this basis.

I do not at this stage wish to pursue this set of theoretical comparisons any further; the arguments would seem to be well-rehearsed. Here I wish to look at some more recent approaches which might be seen as alternatives or complements to the family practices approach. These approaches, again, are often quite different from each other but they all have one feature in common which is to recognise the fluidity of the idea of family and which are less willing to talk about *the* family. These approaches may be grouped, very roughly, under two broad headings. The first set of approaches seems to begin with family relationships or, at the very least, to be much closer to the kinds of debates associated with family studies. These include ideas of intimacy, personal life and configurations. The second set of approaches frequently take, as a point of departure, positions at some distance from family relationships but which come to include them at various points. These include 'caringscapes' and 'the total social organisation of labour'. I shall, in the light of these comparisons, attempt to locate the family practices approach and also to explore how my analysis might be developed or enriched in the light of these comparisons.

Intimacy, personal life and configurations

In this section I look at some approaches which begin with or which are close to family relationships. The authors who have developed these ideas and approaches have been associated in varying ways and degrees with family studies. Apart from that I would not claim more than a loose affinity between these perspectives.

Intimacy

While ideas about intimacy and its transformations under late modernity are chiefly associated with Anthony Giddens' (1992) influential book, I shall take as my point of departure the book on this topic by Lynn Jamieson (1998). Giddens is identified with a particular, relatively optimistic, perspective on intimacy and is largely concerned with adult relationships, chiefly sexual relationships. Jamieson is critical of Giddens' approach and seeks to provide a more general overview, one more informed by empirical and historical research, of the topic. I am less concerned with the debates around Giddens' book but more interested in the extent to which an emphasis on intimacy provides an alternative to or an extension of a family practices approach.

At the outset we should perhaps distinguish between 'intimates', a noun identifying a particular set of relationships which may, perhaps, be contrasted with other relationships such as acquaintances and strangers (Morgan, 2009) and 'intimacy'. As to who counts as an 'intimate' may vary historically and between individuals. Some years ago, when

I taught a course on the subject, students identified the usual suspects such as family, kin, friends, partners and lovers but also argued for pets, deceased family members or friends and even some inanimate objects.

Intimacy, on the other hand, seems to refer to a particular quality of a relationship. Thus even within the sets of people identified in the previous paragraph, some individuals might be defined as not being particularly intimate. While there might be some supposition that relationships between spouses and between parents and children ought to be intimate (in a way to be defined) this more qualitative dimension might not always be present.

This leads to a recognition that intimacy is probably not a one-dimensional phenomenon, but may be understood in different, and not always congruent, ways. Jamieson begins with what she calls 'disclosing intimacy', 'an intimacy of the self rather than an intimacy of the body' (Jamieson, 1998: p. 1). But her argument is that this is not the only way in which intimacy may be understood:

> Close association and privileged knowledge may be aspects of intimacy but perhaps are not sufficient conditions to ensure intimacy as it is more generally understood.
>
> (Jamieson, 1998: p. 8)

'Loving, caring and sharing' may also be seen as dimensions of intimacy (ibid.: p. 9).

I have found it useful, as a point of departure, to distinguish between the following dimensions of intimacy:

(a) Embodied intimacy. This includes but is not limited to sexual intimacy. We can also, for example, include forms of embodied caring under this heading as well as everyday touching.
(b) Emotional intimacy. This is close to what Jamieson is referring to when she talks of 'disclosing intimacy'. This involves sharing and disclosure but might also include what has come to be referred to as 'emotional intelligence' a kind of understanding of the other which is not simply at the verbalised level.
(c) Intimate knowledge. This is partly what emerges out of embodied or emotional intimacy but is more to do with the interweaving of personal biographies over a period, often a considerable period, of time.

To begin to think of dimensions of intimacy is to realise that not all intimate relationships will score highly on all three (indeed, in some cases, embodied sexual intimacy may be prohibited) and that intimacy may sometimes characterise relationships outside the generally understood circles of family, partners and friends (relationships between professionals and clients, for example).

Another distinction which needs to be borne in mind, and one which runs through Jamieson's analysis, is a distinction between discourses or stories about intimacy and actually observed or experienced practices. Thus, the idea that 'disclosing intimacy' represents a desirable goal that all intimate relationships should strive towards, is a particular story that is told about intimacy in late modern times. The relationship between these stories and actual practices is, she argues, a complex one (Jamieson, 1998: pp. 10–14).

On the basis of this discussion of intimacy it can be seen that it would be perfectly possible to extend this in order to speak of 'intimate practices'. To what extent would this provide a more adequate substitute for 'family practices'? To talk of intimate practices, it might be argued would be to provide a perspective that is both more inclusive and even more of a departure from the dangers of talking of a reified 'the family', than using the term 'family practices'. 'Intimate practices' clearly include family relationships but also extend, as Jamieson's book demonstrates, to friends, and non-heterosexual partners.

Thus family practices would seem to be narrower than intimate practices in that there are significant examples of the latter which do not include any strong sense of family. On the other hand, there are some family practices and relationships that could not necessarily be described as intimate. This might apply to some members of the wider kinship network or other extensions through, say, the processes of constituting family relationships through divorce and re-marriage. Some large-scale family events, weddings or funerals for example, may possibly include relationships that could not be defined as intimate on any of the dimensions suggested above while some 'non-family' participants may well be described as intimates. To give another example, the popular activity of documenting one's family tree may well involve the search for others where no intimacies are involved except, perhaps as a result of the investigation, some intimate knowledge.

Further, it is likely that individuals do routinely make distinctions between 'family' and other intimates and that the making of these distinctions is part of what is meant by 'family practices'. They may not necessarily always agree that family relationships are 'above' or 'superior to' or 'take precedence over' these other relationships but they may see them as different and that these differences may sometimes have consequences for the way in which they feel or behave.

To conclude this brief section, there is no doubt that a stress in intimacy encourages readers to think more readily of the broader field of relationships which are both significant and close. Family relationships overlap with, rather than are included within, intimate relationships and both the points of overlap and the points of difference are worthy of investigation.

Personal life

'Personal life', as developed by Carol Smart (2007) has some overlaps with the themes developed in the study of intimacy, chiefly in that both include studies of family relationships but are not limited to these. In Smart's terms, family relationships appear (indeed, they provide most of the illustrations within the book) but they are not automatically centred. She refers to a wide range of family and kinship relationships but also refers to same-sex relationships and, more briefly, to friendships. 'Family practices', on the other hand, are necessarily limited to family or, possibly 'family-like' relationships.

While departing from the family practices approach in terms of the range of relationships considered, she still sees her account as extending my analysis. This is not simply in terms of the relationships that readers encounter in her book but also in terms of the kinds of themes that are dealt with:

> I acknowledge that family is what families do, I also think we need to explore those families and relationships which exist in our imaginings and memories, since these are just as real.
>
> (Smart, 2007: p. 4)

The kinds of themes that she wishes to explore are presented as a set of five overlapping circles labelled 'memory', 'imaginary', 'biography', 'relationality' and 'embeddedness' (ibid.: 37). While these themes can be dealt with from a 'family practices' perspective, it is true that the reader is almost automatically drawn to a slightly different set of concerns. These concerns are both slightly more mysterious but also closer to the way in which personal relationships are actually experienced than the, possibly, slightly more prosaic world of family practices.

Some of the overlaps with and departures from the idea of family practices can be seen in the three pages where she spells out what she sees as distinctive in her treatment of 'personal life' (Smart, 2007: pp. 28–30), a term which she regards as being more than a 'terminological holdall'. If we take her list of eight distinctive features we may explore these overlaps and differences, very briefly:

1. She distinguishes 'personal' from 'individual'. This is partly a departure from liberal and neoliberal political philosophy but also, and more directly, from the 'individualisation thesis' associated with Beck and Beck-Gernsheim (1995, 2002). The chief difference between the two, apparently similar, terms is that she sees the former term as being inherently relational. A sense of the personal and personhood is, she argues, built up through and in relation with others. It will be clear that the family practices approach is also a relational one, although with a narrower frame of reference.

2. She recognises that 'personal life' is an analytical statement in that in seeks some measure of generalisation over a range of relationships and does not necessarily reflect the ways in which individuals understand or talk about their lives. (Normally, one assumes, reference is made to particular relationships such as 'my partner', 'my friends', 'my grandmother' and so on.) This again, is similar to my usage of 'family practices'.

3. 'Personal life' allows for the idea of a life-project but this is distinct from the more free-floating do-it-yourself biographies referred to by Beck and Beck-Gernsheim as well as Giddens. Such an idea is not alien to my notion of family practices but is not directly present in the original statements.

4. 'Personal life' does not prioritise biological kin or marital relationships. This is a clear distinction from the family practices approach although it depends how one defines priorities. To talk of 'family practices' does not necessarily say that they are more important than other sets of relationships and I deal with this issue more fully in my concluding chapter.

5. The concept of personal life, in Smart's usage, conveys a sense of motion. This is more than is conventionally understood by the term 'life-course' and may include all kinds of other movements and transitions, in and out of work, education, relationships and dwellings. Again this is a point of overlap with rather than departure from the family practices approach.

6. The term has a fluidity about it and transcends conventional distinction such as those between the public and the private. There are some points of contact here with my discussion of family practices.

7. Personal life 'also gives recognition to those areas of life which used to be slightly below the sociological radar' (ibid.: p. 29). I have already drawn attention to this aspect of Smart's approach and it is one to which I am greatly sympathetic. I hope this sympathy will be apparent in the pages to come.

8. Personal life departs from the standard model of the heterosexual, white, middle-class family which she sees, probably correctly, as dominating much discussion of family life. I suspect that I might also be guilty of this and, again, this is a theme that I shall return to in a later chapter.

Thus, as Smart states quite explicitly, there is probably more evidence of continuity and overlap between her approach to personal life and my approach to family practices than there is evidence of rupture. From my perspective, I find that many of the themes developed in her book can be applied, with some profit, to the family practices approach. For an example, consider her approach to 'embeddedness' (Smart, 2007: pp. 43–5). Following on from this idea it is possible to conceive of individuals' family practices as varying in density. Some people wear their family practices lightly as a kind of accessory rather than as an essential garment. Others are clearly distinguished by their family practices. This variation, of course, may not be simply between individuals but may vary in the course of a life. The differing density of family practices does not simply depend upon the frequency and number of such practices but also on the extent to which they are carried out with reference to other identified family members. Smart usefully points out that this embeddedness or density can be both a source of ontological security and/or a source of constraint and oppression.

Where the kind of approach outlined by Smart and my family practices approach can clearly overlap is in the aim to bring out hitherto under-explored or unexplored areas of family and inter-personal life such as material objects, memories and secrets. She seeks to 'reflect complexity and ambiguity without being confusing and incomprehensible' (Smart, 2007: p. 186). Undoubtedly, her explorations in this book (and, as she suggests, possible future volumes) achieve this and produce some fascinating insights and I hope this stimulus will be reflected in this present volume.

So, it might be asked, why not abandon 'family practices' in favour of 'personal life'? I hope to return to this question in the concluding chapter but at this point I will state that ideas and experiences of family are still seen as important and distinct by significant numbers of people and that there are some distinct research problems thrown up as a consequence of using family as a point of departure. This, of course, makes no presumption of functionality or centrality from either a societal or an individual perspective. From my own perspective there seem to be good reasons (other than simple inertia) for continuing to look at family practices and to see how they overlap with other areas of social life. But, in continuing to be concerned with family practices, my explorations will be greatly enriched by thinking more generally about personal life.

Configurations

The configurational approach, identified in particular with the work of Eric. D. Widmer (Widmer and Jallinoja, 2008), begins at what is by now a familiar point of departure, one common with the approaches already discussed as well as with the ideas of family practices. This is the desire to go beyond reference to the household or to a 'limited set of family roles' and to question the 'emphasis on the nuclear family' (Widmer et al., 2008: p. 1). The title of the book edited by Widmer and Jallinoja (2008) is *Beyond the Nuclear Family: Families in a Configurational Perspective* and this sense of 'going beyond' is what is found in all the approaches mentioned so far.

We also find another theme, common to the ideas associated with 'intimacy' and 'personal life', and that is a questioning of the individualisation thesis and the strong emphasis on the self and individual projects to be found in accounts of intimate relations in late modernity, whether these accounts be pessimistic or optimistic. In so doing, the studies collected in this book and elsewhere have affinities with approaches that talk about social networks and social capital and all accounts that attempt to develop or preserve a more relational approach to social life in a late or postmodern world. Some of the sources cited by Widmer and his colleagues include Moreno, Lewin and Elias (Widmer et al., 2008: p. 5).

We also find, in these preliminary statements, a critical point of departure from my family practices approach. Concepts like 'family practices': 'do not convey the fact that family relationships, because of their complexity, are likely to remain highly patterned and embedded in the social structure of late modernity' (Widmer et al., 2008: 3). I shall return to criticisms of the family practices approach in a later chapter. Here I point out that Widmer and his colleagues, partly through their stress on the importance of structural constraints, are indicating some points of difference from the family practices approach.

In principle, the configurational approach need not deal with family or family type relationships and could deal with any chains of relationships and inter-dependencies in society. This certainly seemed to be implied in some of the identified founders of this approach such as Elias and Moreno. In practice, however, the studies presented by Widmer and Jallinoja do not stray greatly from discussions of family and kinship. Thus one definition of configurations runs, 'sets of directly or indirectly interdependent persons sharing feelings of family belonging and connectedness' (Widmer et al., 2008: p. 3). The emphasis is on *family* configurations.

Four key ideas are identified as constituting the (family) configurations approach:

1. Families should not be defined by institutional criteria. Instead we need to focus our attention on actualised relationships.
2. We should not focus on specific dyadic relationships (parent-child, husband-wife, brother-sister) but rather should always be aware of the ways in which these are embedded in larger networks of relationships.
3. Individual and group structures are interconnected, and the one is not to be subsumed by the other.
4. The historical and spatial nature of configurations should always be remembered; 'all human configurations evolve through time and place'. (Widmer et al., 2008: p. 6)

On this basis, Widmer and Jallinoja present a rich variety of case studies illustrating the configurational approach and demonstrating its utility. These include an account of the importance of aunts and uncles, discussion of children's perspectives on family configurations, post-divorce family relationships, migration and family members undergoing psychotherapy. Issues of gender run through nearly all of the studies, and caring relationships feature strongly in many or most of the studies.

A good illustration of what is entailed in a configurational approach is provided in this quotation from a study of caring for dependent elderly parents:

The husband can also, as in the case of Beatrice, be a go-between. Beatrice's father is the main carer of his wife, who suffers from Alzheimer's disease. His two daughters have had some difficulty in persuading him to make use of professional help to give him some respite. The systematic opposition of their father is sometimes a source of exhaustion for Beatrice. Her husband, who works in the geriatric sector, plays an important role for her father. Her father listens to him and often follows his advice, much more than he does with is daughters.

(Le Bihan and Martin, 2008: p. 66)

While it might be possible to break this account down into a set of overlapping dyads – sister-sister, parent-child, husband-wife, formal-informal and so on – it is clear that this 'whole', however fluid and loosely bounded, is greater than the sum of its constituent parts.

Some of the rich complexities that emerge from this approach emerge in a Portuguese study of men's family configurations (Wall, Aboim & Marinho, 2008). They identify seven different patterns in men's configurations which are made up of different combinations of partners, siblings, other kin and friends. The emphasis on men is a useful corrective to some more stereotypical understanding of gender differences in personal relationships but the most important point is the way in which the authors stress the importance of a life-course perspective. The ties that they describe in some detail are ties that have built up over time, through the men's involvement in different households, different locations and different educational and working experiences.

Something of a similar idea is conveyed in a discussion of the concept of *entourage* which is translated as 'contact circle' (Bonvalet & Lelièvre, 2008). These are networks of key figures, in an individual's life, and their relatives; these key figures may also include non-relatives. As in the Portuguese study, the authors stress the importance of past history as well as current ties.

We can see that there are several points of contact between this con-figurational approach (including the idea of the *entourage*) and other approaches identified in this chapter. Chief among these is the desire to move 'beyond' the nuclear family. This movement is in two senses. In the first place it moves beyond the institutional models of the nuclear family or household in which these relatively bounded entities are seen as building blocks within a wider social order. In the second place, they move beyond the relationships which are conventionally seen as being contained within these entities (parents and children, partners) to con-sider wider sets of kin and, sometimes, non-kin relationships. This also combines with a desire to focus on actualised relationships rather than more abstractly identified ties and here one may see some overlaps with a family practices approach. Thus a family tree may identify all kinds of family connections but we need other kinds of data to see the extent and the ways in which these connections are actualised in everyday practices.

In one sense this departure is less radical than it may seem at first. Even the most conventional nuclear-family-based household (as Parsons recognised when he wrote of the kinship system of the US) is linked to other households by virtue of the fact that the parents within this unit themselves have or had parents and siblings and that, therefore, links across generations are built into the nuclear family model. It is rather like those picture puzzles that may at first be seen one way and then, later, as something quite different. The configuration approach recognises that the overall picture may be more ragged and less symmetrical and that it may be affected by transitions and trajectories other than moving out of the parental home in order to form a new nuclear-family-based household but, ultimately, it represents a development out of rather than a substitute for earlier approaches.

Linked to this question is the extent to which this approach represents a significant departure from approaches in terms of social networks (or indeed from earlier influential studies of kinship), studies which are referred to at different points in these studies (Phillipson, Allan & Morgan, 2004; Spencer & Pahl, 2006). Certainly many of the concerns developed in network studies (social capital, for example) are present here and much of the analysis would seem to have some strong affinities with the networks approach, affinities which the authors recognise. The only difference I can find, and again this is a difference of emphasis, is the emphasis upon time and history and the way in which these ties develop over the years in ways which are distinct from more conventional life-course models. A further difference would seem to be that the configuration approach is not simply added on to or coexisting with other forms of family analysis but should be seen as a substitute for these. A strong version of configurational analysis (although in fact the argument rarely, if ever, goes this far) would be to argue for the substitution of 'configuration' for 'family'. Such an approach, which I stress is only hinted at in these pages, would be a valuable thought experiment but would probably not have a great appeal in public policy or everyday talk.

A core concern with Widmer and his colleagues is the idea of structural constraints, something that they feel is under-recognised in some current discussions of family and intimate life including the family practices approach. Family configurations, therefore, do not simply emerge in some free-floating fashion but are actualised responses to real external pressures:

> Neither fully predictable nor random, family configurations are patterned responses to the complex set of sociological and psychological constraints stemming from post-modern life.
>
> (Widmer & Sapin, 2008: p. 301)

In fact the connections with 'post-modern life' are not always clear in the studies presented (the onset of Alzheimer's disease, for example) but the point about sociological and psychological constraints is well taken. Here the configurations become actualised as family members attempt, in collaboration with others, to respond to these constraints and crises.

But to what extent do these configurations themselves constitute constraints? For example, Finch and Mason (1993) show how, in the course of time, individuals within sets of family relationships develop particular moral reputations which they feel that they have to live up to. Sets of reputations and obligations build up over time, perhaps initially in response to a particular family or external crisis, and provide a framework of constraints in later years. Family configurations constitute a form of habitus which can act as a constraint when times or situations change.

Nevertheless it can be seen that the configurational approach represents a valuable addition to the tools for family and relational analysis. That this approach can be seen as overlapping with some other approaches discussed here is not necessarily a sign of weakness but an indication of further work to be done.

Other approaches

The approaches outlined in the previous section have at least one thing in common. While they all talk, in different ways, about 'going beyond' the family, they all have some affinities with family relationships. Thus family relationships are included within or overlap with these other approaches or, as in the case of the configurational approach, seek to extend family relationships through chains of kin or friends.

In the approaches indicated in this section, the point of departure is, to a greater extent, 'outside' the family. As we shall see, this is a matter of degree, but the distinction is an important one. One of the features of 'family practices' approach is that of the fluidity of family boundaries and the ways in which certain practices may be viewed as 'family' from one perspective and as something else, 'class' or 'gender' for example, from another. Thus we may start by looking at family relationships and end by exploring other areas of social life. Similarly, we may begin somewhere else, food, leisure or transport for example, and find ourselves considering family relationships (Morgan, 1996). Here I shall look at 'caringscapes' and, 'the total social organisation of labour'.

Caringscapes

One of the articles which introduces the 'caringscapes' perspective (McKie, Gregory & Bowlby, 2002) begins with the gendered interface between home and work and the practices adopted by mothers (especially) in order to care for children or other family members while maintaining paid employment. 'Care', with all its different connotations, has increasingly been a focus of scholarship over a wide range of disciplines and has come to be seen as a fundamental human activity. While, in modern and late modern societies, the practices of care have been located within 'the family' this identification is strongly shaped by ideological and political considerations. Care is also strongly gendered in that it is particularly identified with women. Care needs of individuals within family networks change over time and, while often identified with families, can take place in a variety of social spaces.

Theoretically, the 'caringscapes' perspective begins with the notion of 'timescapes' developed by Barbara Adam (Adam, 1995; McKie, Gregory & Bowlby, 2002). This orientation is brought to bear in the exploration of time-space relationships in the practices of caring. Also fed into the argument are considerations of power relationships and social inequalities and distinctions between the public and the private.

Both 'caringscapes' and 'timescapes' (and also, later, 'healthscapes', McKie, Bowlby & Gregory, 2004) are metaphorical expressions originating in the more familiar word, 'landscapes'. Perceptions of landscapes depend upon the standpoint of the observer or participant. Notions of height or distance also involve questions of time ('a long way away' may be expressed in terms of walking or driving) and human projects: '… that mountain is only steep if I want to climb it or if climbing mountains is a possible project for human beings' (Craib, 1976: p. 31).

It is not difficult to move from this to everyday questions of care. An employed mother has to consider the timetables of other people (neighbours, kin, paid child-minders) or institutions (workplaces, schools, nurseries) and the distance of these other people or institutions from home. Distance, again, is not simply a question of miles but of the time taken and this may reflect factors such as traffic congestion or bus timetables. Further, all this exists in a longer temporal framework involving the life-courses of the mother concerned and of others connected to her. This whole 'caringscapes' perspective is neatly encapsulated in two sentences:

A caringscapes perspective reflects the range of activities, feelings and positions in parents' or carers' mapping and shaping of caring and working.

Caringscapes include past and current experiences, knowledge of the experiences of others and anticipation about the changing nature of caring and work.

(CRFR, 2004: page 1)

Important here is the way in which 'activities, feelings and positions' are linked to each other and the reference to the interweaving of past and present.

Later, these ideas are extended or adapted to the idea of 'healthscapes' which raises slightly different, but clearly overlapping, considerations: 'Much health work involves scheduling and working across locales, in particular mediation with health professionals and access to health services' (McKie, Bowlby and Gregory, 2004: p. 603).

From the perspective of this present work, the 'caringscapes' perspective is significant in a variety of ways. In the first place, and in common with some of the other approaches, it provides a way of decentring the family. The points of departure are issues to do with health or care rather than 'the' family. Family relationships are clearly implicated, indeed often central, within this framework but their fluidity is demonstrated by the focus on time and space. Second, this approach does provide a ready way in which issues of structured inequalities and power may be incorporated. To return to the metaphor, these considerations contribute to the steepness or distance of key features of a given landscape. Finally, this approach provides links between individuals and their life trajectories, social configurations and social institutions all located within a temporal framework. At the very least, the 'caringscapes' perspective provides a framework within which it is possible to gain a deeper understanding of the working and meaning of family practices.

The Total Social Organisation of Labour

At one glance the idea of 'total social organisation of labour' (TSOL) (a term that its originator now admits could have been less clumsy) is at a considerable distance from family practices. But therein lies its importance. It provides a good demonstration of how quite a different point of departure can quickly implicate family life and practices.

An early statement of the key ideas can be found in a study of women in British manufacturing industries between the wars (Glucksmann, 1990). This study refers to a set of overlapping debates that frequently entered into feminist and socialist writings; the relationships between capitalism and patriarchy, between class and gender and between market and domestic economies. Here, Glucksmann begins to develop the idea of TSOL as a way of understanding and, to some extent, transcending these dichotomies or polarities.

The first point to make about TSOL is that it is an open and evolving topic. This is partly because the social landscape (including the nature of work and households) that it attempts to describe is itself changing and that, therefore, the term should allow for these changes. Further, the aim is not so much to build up rigid theoretical models of a structuralist or a functional kind but to develop a flexible way of understanding a complex reality. At the same time a key concern remains with class and gender inequalities, inequalities which increasingly take on a global significance.

A point of departure was the various challenges to the idea that the term 'work' could be equated with paid employment. This limited understanding was challenged by feminist scholars concerned with domestic labour, an increasing interest in emotional or aesthetic labour as well as some more specific and influential studies such as Pahl's *Divisions of Labour* which considered a whole range of paid and unpaid forms of labour (Pahl, 1984). Two quotations underline the importance of these insights and the way in which they lead to some of the key ideas of TSOL:

> Work is not assumed to be a discrete activity carried out for remuneration in institutions (although it can be) but, rather, is conceptualised as being embedded in other domains and entangled in other sets of social relations.
>
> (Parry et al., 2005: p. 4)

> [A] significant proportion of labour in advanced industrial societies may also remain undifferentiated from informal, household, familial or community relationships which contain other components in addition to work.
>
> (Glucksmann, 2005: pp. 31–2)

An example provided here refers to the care of children, a necessary activity that, as was emphasised in the previous discussion of 'caring-scapes', links and cuts across workplaces, households, state institutions and communities.

Throughout the various workings and re-workings of this concept, it is stressed that this is not to be treated as a straitjacketed, constraining social enquiry but as a flexible tool seeking to understand processes, exchanges and movements. It seeks 'to analyse the patterns of exchanges, reciprocities, dependencies and inequalities that straddle the multiplicity of forms and relations of work' (Glucksmann, 2000: p. 53). Rather than explore all these individual terms in detail, a sense of what is entailed in TSOL approach may emerge from some more specific examples:

- The growing interest in forms of emotional and aesthetic labour inevitably cannot confine itself to particular workplaces but involve a consideration of a range of sites (paid employment, domestic, leisure based) where these forms of labour are practiced and the relationships between these different sites.
- In the course of one of the earliest statements of the TSOL, Glucksmann notes how working-class women, employed in making a novel range of consumable durables between the wars, also came to perform work as consumers, buying those self-same products in order to build and to rebuild homes and the idea of home (Glucksmann, 1995). Consumption is increasingly to be seen as a form of unpaid work, again linking various sites.
- The process of getting a job in a particular office or factory may have been, certainly in the past, assisted through networks of family or neighbourly relations.
- An increasing stress on 'flexibility' and 'family friendly policies' by employers or governments explicitly recognise the complex connections between home and work.
- Parents are increasingly, willingly or otherwise, drawn into their children's education. This entails a blurring of the boundaries between family and education and, often as in the case of working mothers, between home and work. Individuals as parents are, in this and in other cases, drawn into the work of others (Reay, 2005).

These examples could be multiplied but it is hoped that they serve to illustrate the kind of terrain where the TSOL operates.

One point stressed by Glucksmann is that the idea of TSOL can be deployed at any level of generality (Glucksmann, 2000: p. 162). Perhaps the easiest way of understanding is in terms of the individual moving through different sites over a period of time. The methodological device of the 'time diary' in which informants are requested to note down their daily activities in some detail may provide clear indications of these movements across different sites and engagements in different sets of social relationships. Such accounts stress that more analytical distinctions between home, work, leisure etc. are, at the very least, blurred in the course of everyday practices.

At the same time we may examine these connections in more macro terms, exploring the historical links between particular places of employment, particular communities or residential estates and particular forms of household. These accounts can, and increasingly should, take on a global character as the connections and articulations cross national boundaries. Yet again, the researcher may operate at a level somewhere between the global and the individual, looking at particular communities or residential areas. TSOL, therefore, also overcomes some conventional analytical divisions.

Clearly, the TSOL is much wider than the idea of 'family practices' but this is not to say that it is irrelevant to these practices. Indeed, it provides one way of showing how family practices can be linked to or seen as other kinds of practices. Further, the TSOL and family practices address somewhat similar problems, departing from a dissatisfaction with the deployment of conventional terminology ('work', 'family') while wishing to acknowledge the abiding significance of the practices subsumed under these various terms.

We may see some particular connections between some of the other ideas outlined in this chapter and TSOL. Some links with 'caringscapes' have already been mentioned and there are some affinities with the 'configurational' approached discussed earlier. Indeed, Glucksmann uses the word configuration at one point (Glucksmann, 2000: p. 163) although her usage seems to be somewhat broader than its usage by Widmer and others (Widmer & Jallinoja, 2008). Arguably, Glucksmann is closer to the earlier use of the term by Elias.

Perhaps the main point of contact between the TSOL, family practices and some of these other ideas is that we are dealing with relationships and with processes. Indeed, Glucksmann (2000, p. 163) uses the term 'relationality' when she is also writing about configurations. The focus of attention is less upon the particular points or individuals that might appear in social analysis (place of work, home, manager, mother) and on the relationships and connections between them. And these relationships are to be understood in processual terms. Glucksmann (2005) makes it clear that TSOL is an evolving term and that new problems (how do you distinguish between work and non-work, for example?) may emerge in the course of the moving analysis. But it is in the clear demonstration of the blurring of institutional boundaries that I find the closest affinities with the idea of family practices.

Conclusion

In this chapter I have attempted to locate the idea of family practices by considering it in relation to other theoretical orientations which might be seen as alternatives to or competing with the practices approach. Clearly, this chapter does not represent a comprehensive overview of all existing or possible theoretical approaches to the family such as rational choice theory, exchange theory or the many versions of systems theorising. Overviews of these, and other approaches, to the study of family life may be found elsewhere.

The aims here are somewhat more limited in that I am primarily concerned with approaches which might already seem to have some kind of affinity with the practices approach or which I have discovered serendipitously in looking at slight different sets of issues. Some of these other approaches refer directly to family practices while others do not. Nevertheless there does appear to be a broad affinity between all these cited approaches.

In concluding this chapter I want to avoid two extremes. In the first place I do not wish to treat these as competing players in some kind of contest in which the practices approach emerges triumphant. But, second, I do not wish to bind all these approaches together into some grand synthesis. I want to preserve the distinctiveness of these and of other approaches that do not appear in this chapter. Nevertheless, and not surprisingly, we can see some common themes and concerns emerging.

The first commonality is the clear desire to go 'beyond' the family. This is, in part, a desire to go beyond what has been called 'the standard North American family' or SNAF (Smith, 1993) which reflects more an ideological construction than a reflection of lived experiences. (Clearly, these concerns are not limited to North America as is demonstrated in the texts cited here.) While elements in this model may vary, core points of focus will include heterosexuality, the combination of parents and children and a relatively bounded and identifiable entity located in a home. There is also the implication that this combination of relatively stable elements carries with it some normative force.

Even staying within the limits of this paradigm it is possible to elaborate a somewhat broader picture. This could be done through adopting a life course perspective which shows how family members form different clusterings in different household structures as they move through their individual life trajectories. It can also be done by simply recognising that individuals (e.g. two heterosexual parents) are inevitably linked to other households through the simple fact that they themselves have parents or siblings. Hence it is possible to go 'beyond' the standard model without moving very far from base.

The 'configurations' approach outlined here tends to treat the 'beyond' in this way although hints at, and sometimes deals with, other extensions as well. But the frame of reference in the approaches identified with intimacy and personal life begin from a much wider range of relationships and clusterings that individuals might find themselves in at different times. These include friendships, non-heterosexual partnerships and, possibly, community relationships. Within these, family ties are important but are not to be identified with the whole.

Another sense of the 'beyond' is conveyed in the 'caringscapes' and the TSOL approaches. Here, the concern is more with establishing a framework that can explore a whole range of linkages between practices, individuals and different agencies and institutions. Family relations and practices are implicated within a much wider framework to do with the ways in which care is organised and delivered and the linkages between all different kinds of work in a modern society. Family and household relationships are seen as being key elements but within a framework that does not refer directly to intimate or domestic life. These are approaches which are more fluid and more dynamic than some earlier functional or systems approaches which also, in some measure, addressed similar problems.

In contrast to the approaches outlined in this chapter, the family practices orientation might seem to be somewhat narrower, more limited, in focus. Insofar as this approach goes 'beyond' anything it is in terms of surpassing frameworks based upon relative static models of *the* family, frequently identified with the nuclear-family based household. This would seem to be something less than what appears to be offered by these alternative approaches.

I shall return to some of these questions in the next and in the concluding chapter. Broadly speaking, I recognise that the family practices approach is, in some way, a limited approach but one which highlights problems which are not centrally considered in any existing alternatives. These include questions of inter-generational relationships and their connections with issues of historical time and memory. Further, I should argue that the relative fluidity of the practices approach allows for an exploration of the points where family practices overlap with, articulate with and meet other sets of practices. From different perspectives, I argue, the same set of practices can be seen as family practices, class practices, gender practices or working practices. This also underlines the fact that family practices can be seen to be taking place away from the sites conventionally associated with these practices. Thus the male employee who takes on extra paid work in order to provide for a newly arrived child can be seen as engaging in family, as well as working, practices (Miller, 2010). Thus the practices approach can also be seen as going 'beyond' conventional boundaries and understandings but in different ways.

I have argued that all the approaches identified here (as well as, doubtless, others) are attempting to go beyond pre-ordered categories and boundaries but that they do so in different ways. Another shared feature is the desire to deal with fluidity and complexity in modern life. The term 'complexity' is itself not a straightforward one but the following elements seem to be important:

- Fluidity of boundaries. This is a feature shared by all these approaches although, again, this sense of fluidity may be expressed in different ways. Notions of personal life, for example, have almost a built-in sense of flow as this life is built up and reshaped on a day-to-day basis. Family practices can be seen as other kinds of practices and TSOL can be seen at all kinds of different levels from the individual to the global.
- A combination of relationships which are both multiple and multiplex. It is a truism to say that any one individual is linked to, interacts with, numerous other individuals in the course of a day or a lifetime. Further these others are located in different spheres of life such as family, employment, community and so on. In addition some, but not all, of these relationships are multiplex in that the others to whom one is connected are also, in different ways and different degrees, connected to each other. Thus there are clusters of others within family constellations, places of employment and neighbourhoods. Much of the complexity of modern life arises out of these different kinds of relationships and the ways in which they are articulated, not always harmoniously.
- The continued presence of uncertainty, risk and unpredictability. While none of the approaches outlined here deal with these issues

directly, they remain waiting in the wings and give these different ways of describing social relationships their significance and their poignancy. Thus, family practices although so frequently associated with the routine and the repeated also necessarily are confronted with mortality, sickness and cruelty. The same is true of personal and intimate relationships. Caringscapes are in part organised around these uncertainties and calamities just as TSOL has to deal with the unpredictabilities of local and global economies.

This list is, in keeping with the very idea of complexity, not complete. But it should provide some sense of the core concerns which link these different ideas. These ideas provide a set of tools which can be used in all kinds of different combinations or which can be discarded if they are seen to fail in their task of confronting the complexities of modern life.

4
Developments and Difficulties

Introduction

The previous chapter looked at some approaches to family and intimate life that might be seen as alternatives to the practices approach. This chapter focuses more specifically upon family practices, the ways in which the concept has been used and some possible criticisms of the idea. Since I first outlined what I understood by family practices in 1996, several people have referred to, used and sometimes developed the concept. The first section of this chapter will look at some of these developments including, as a more extended illustration, a British study by Julie Seymour (2007). The second section will look at one particular extension, a going beyond, of the idea of family practices that looks at the idea of 'family display' (Finch, 2007). In the third section I shall look at some possible criticisms of the idea of family practices and the ways in which considerations of these criticisms might be used to develop the concept further.

Usages and developments of 'Family Practices'

As noted above, since I elaborated the idea of family practices in 1996, several researchers have referred to, used and sometimes built upon these ideas. These references range from simple and unelaborated citation through accounts which highlight one or two of my key points and on to somewhat more elaborated discussions. I do not claim a systematic trawl of all the references to 'family practices' here. My choice to some extent reflects my own academic and friendship networks just as the citations to my work reflect, in many cases, similar and overlapping networks. The aim is to show the range of family-based issues that have been addressed using or, at least, referring to 'family practices' and to suggest those features of my original presentation that appear to have attracted most attention.

'Family practices' before 1996

I have argued that the notion of 'practices' was already in widespread use by the time I came to explore how this term might be used in a family context. I attempted not only to adapt this term to family analysis but also to elaborate what was entailed in this move. I did not know at the time that the actual term 'family practices' had already been used.

This usage occurred in a study linking elite family practices with state formation in the early modern Netherlands (Adams, 1994). The term was used but not defined or elaborated and no source was given so that one can assume it was a simple adaptation of a term that was already familiar in scholarly circles. In terms of the usage of the term in this particular article, 'family practices' could be roughly equated with 'strategies' and a reference to Bourdieu on marriage strategies reinforces this impression. Further, in the article it is also clear that family practices were also gendered practices.

While being slightly disappointed to learn conclusively that mine was not the first usage of the term 'family practices', I consider this discovery to be important in two ways. First, it reinforces my sense that the term 'practices' was very much in use around this time and that it would not be unexpected for the term to be qualified by the word 'family' eventually. The fact that the term is not presented in quotation marks suggests a degree of naturalisation that should perhaps not be unexpected. The second point of significance is the use of this term in a historical context and in an elite context where family practices would almost inevitably overlap with political and class practices. This raises an issue for further investigation, that is historical and cultural differences in family practices, differences which revolve around the degree of overlap between these practices and other areas of social, political and economic life.

Under-elaborated usages

I have said that the term 'family practices' was used but not elaborated in this 1994 study and this is true for quite a few studies published after 1996. Thus Oinonen (2008) has a chapter called 'On Family Practices' which, as in the 1994 article, takes the term for granted as being one that is in current usage. The chapter contains demographic data contrasting indicators such as marriage and divorce rates, proportion of cohabiting couples and fertility rates for Finland and Spain. Here the usage is one which contrasts 'practices' with 'family values' and 'family ideologies'. A similar example where the usage emerges from implied contrasts is to be found in a study of Swedish family policy (Ahlberg, Roman & Duncan, 2008). Here the contrast is underlined in the title of the article: 'Swedish policy rhetoric versus family practices' and there is also an implied contrast between theory (ideas of individualisation and 'the democratic family') and practice. These, of course, as was demonstrated in Chapter 2, reflect some more everyday usages of the term 'practice'. Again, the term is used but not elaborated.

Another usage, closer to the idea of 'strategy', is to be found in a study of middle-class child-care arrangements (Vincent & Ball, 2006). This refers to the 'class and gendered nature of 'family practices'' (Vincent & Ball, 2006: p. 2) thereby highlighting the fluidity of the idea of practices. Child-care arrangements, themselves highly gendered, can be linked to processes of class reproduction. Again this is an example of a use of the *term* rather than the *concept*.

An apparently popular feature of the idea of family practices lies in its emphasis on doing rather than being and the way in which this highlights a sense of agency. This is evident in Gatrell's study of parenting (2005) where she also highlights the way in which the idea of practices enables readers to move away from constructions of *the* family (Gatrell, 2005: pp. 37–44). A similar emphasis on 'doing' is to be found in Sweeting and Seaman's argument that this serves as a useful corrective to the 'tick-box' approach where respondents are asked to respond to 'who do you live with?' The context is one which examines the boundaries of children's perceptions of social networks.

These references, where the term 'family practices' may be used but barely elaborated, could be multiplied. The point here is not to call these studies into account for some failure or oversight. It is rather that these studies reinforce the argument that the idea of 'family practices' has increasingly gained currency and may be deployed much in the same way that previous researchers might have used terms such as 'role' or 'function', that is without further elaboration. In such cases, the meanings of the term lie in the usages and that these usages, not surprisingly perhaps, can be seen to reflect the range of usages to be found in a large dictionary.

More elaborated usages

One of the earliest usages is to be found in Smart and Neale's study of post-divorce parenthood, *Family Fragments?* (1999). They re-emphasise the difficulties involved in talking about *the* family and welcome the way in which the practices approach departs from treating the family as a thing and moves towards an emphasis on fluidity and overlap with other practices. In conducting their empirical study they find it helpful to combine a practices approach with Finch and Mason's emphasis on negotiation (1993).

Smart and Neale stress that the practices approach implies ideas of agency. Interestingly, and this is not something that I explored in any depth in my original statement, they see this as a way of overcoming the agency/structure divide:

> In using the idea of practice one automatically imports with it the idea of the actor or the agent. So one ceases to pose structure and agency as discrete entities. Practices require agents to carry them out, so to speak. But because they are fairly routine and located in culture, history and personal biography, they are not free-floating, random or serendipitous.
>
> (Smart & Neale, 1999: p. 73)

The particular application elaborated in this study is an exploration of families undergoing transition as a result of divorce or separation and with a particular emphasis on post-separation parenting arrangements. They see the practices approach (among other linked approaches) as one that avoids the 'social problems framework' that has so often char-acterised discussions of these issues (ibid.: p. 40). Whether engaging in co-parenting, custodial parenting or solo parenting and whether formally present or absent, the participants and their children are all engaged in family practices. It can be argued that there is an affinity between the fluid circumstances of the family members undergoing these transitions and the relative openness of the practices approach.

Smart and Neale found the practices approach useful in thinking about family situations that departed from a dominant normative model and we find something similar in Weeks, Heaphy and Donovan's analysis of same sex intimacies (2001). The subtitle provides a clear indi-cation of the scope and orientation of the study: *Families of Choice and Other Life Experiments*. As with other scholars, they welcome the way in which the practices approach emphasises doing

> a social scientific approach, which may not resonate easily with peo-ple's experiences, but it has an important implication for understand-ing non-heterosexual lives. It displaces the idea of family as a fixed and

timeless entity ... We see it instead as a series of practical, everyday activities which we live: through such tasks as mutual care, the division of labour in the home, looking after dependents and 'relations'.

(Weeks, Heaphy & Donovan, 2001: p. 37)

They also take up the idea that family practices need not necessarily be confined to the home.

These researchers (as the quotation above suggests) are keen to explore meanings in action rather than to impose definitions on to sets of research subjects. As might be expected, they found considerable variation both in understanding of the term 'family' and in assessing the term's value. Some defined 'family' in terms of a particular quality of relationship and one that could easily be applied to friends as to blood or affinal ties. Others saw the terms 'family' and 'parenting' as co-extensive (ibid.: p. 160) and the authors themselves refer to caring and parenting practices. Others reject the loaded term 'family' 'while acknowledging the reality; the existence of a complex vocabulary of values which act as the moral cement for non-heterosexual lives' (ibid.: p. 76). Given this fluidity of understanding and evaluation, the practices approach might seem especially appropriate.

Issues of everyday heterosexuality and how it is reproduced over generations are explored in a study by Hockey, Meah and Robinson (2007; Meah, Hockey & Robinson, 2008) in a study of the construction of heterosexuality over three generations. Issues of sexuality have frequently been explored apart from family studies so that an important feature of this study is the way in which the two are brought together. There are several references to family practices in the research reports and the themes that are highlighted deal with the overlap between family practices, gender, sexuality and generation and, linked to this, the interplay between history and biography.

A quotation from one of their respondents, Sarah (aged 43 at the time of interview and a member of the middle generation) may illustrate this:

A generation earlier, Sarah also experienced an upbringing where sex was not talked about. When she naively referred to Tampax over a meal, her father went *very very purple in the face* and her mother, Joan, said *We don't talk about that at the tea table*. He was *quite a Victorian*, strict with his children yet involved in secret long term affairs.

(Meah, Hockey & Robinson, 2008: p. 466)

At one level we can see this as an example of routine family practices; a mealtime in which parents and children get together, during which one of the children is told in strong terms (verbally and visually) what are inappropriate topics for the tea table. Gender and generational differences are clearly being reproduced. But the announcement of a taboo also points to the overlap between family and (hetero)sexual practices. Sex is a mystery which is very much bound up with emerging gendered identities. Later generations (as the study shows) might take part in much more explicit discussions of sexuality but the effect, the construction of gendered, heterosexual identities, will be very much the same.

One final recent example comes from a study of South Asian Muslim families living in England (Becher, 2008). Here, the word 'practices' is used in the title and there is a fairly extensive outline of the key ideas in the first chapter (Becher, 2008: pp. 4–5). The author also uses the term 'parenting practices', a key concern of the book. One of the most important features of the study is the way in which it links everyday parenting practices with issues of ethnic and religious identity; in fact the term is sometimes extended to 'religious family practices'. Within this framework a whole range of everyday family practices such as deciding on a child's education, food and consumption, dress, language and watching TV are given a particular, if often complex, unity within an overarching frame of meaning. This does not mean a uniformity of practices or evaluation of these practices and the author points to several points of tension and variation. One interesting example shows how practices may sometimes involve not doing something as where a father discourages talk about work or school at the dinner table (Becher, 2008: p. 62). Such an informal prohibition reinforces the idea that family life is something distinct from everyday life and again shows a particular way of constructing ethnic boundaries through everyday practices. Again, and further, the author reminds us of the wider context of Islamophobia that also impacts on these practices and their significance.

I could provide several other examples of the uses of the family practices in a diversity of research contexts. These could include food practices (Curtis, James & Ellis, 2009; Metcalfe et al., 2009), family and social change in Wales (Charles, Davies & Harris, 2008), later-life families (Chambers, Allan, Phillipson & Ray, 2009) and fatherhood (Collier & Sheldon, 2008; Dermott, 2008; Miller, 2010). Rather than go through these, and other usages, I shall explore one further example in some more detail.

An extended case: 'Treating the hotel like a home'

An extensive use of the family practices approach is found in an article by Julie Seymour who deals with the kind of atypical cases that can illuminate wider issues (Seymour, 2007). Numerous studies have discussed the relationships between home and work but they have been relatively few detailed accounts of sites where the home is also the place of work, in this case, hotels, pubs and boarding houses. An even more complex example may be provided where businesses such as farms may also be engaged in the hospitality trade.

Various strands of the family practices approach are brought to the fore in this article. The idea of 'doing' family is stressed as we see how family members reflexively create and reproduce their identities and relationships through attempting to demarcate domestic and business time and space. The overlap between family and other, in this case working, practices is explored in some detail. The focus is on 'constructing and protecting family space' on the one hand and marking out business space on the other. Often, clearly marked doors separate the two although there may be some fluidity as, for example, domestic bedrooms might be put to use during busy seasons. What is important is that family members are required to think about and to act upon what is frequently taken for granted; the borderlines between home and work. In engaging in these boundary strategies, family and domestic identities are reproduced.

Again, a quotation may illustrate these complexities:

> Now ... at times like family birthdays ... Jack [*guest during week for more than 4 years*] or any long-term guest would be invited to join us in the dining room and he would come through and help wash up. But other than that, we've always been able to sort of keep family and guests separate. (Interview 15, female boarding house owner)
> (Seymour, 2007: p. 1104, italics in original)

Here the nature and importance of these demarcations is demonstrated on those occasions where some breach, clearly recognised as such, is permitted. Birthdays are frequently occasions for family practices but here there is the added dimension where work spaces adjoin domestic spaces. We can also see an example of family display here (see next section).

This article does not simply deploy the idea of family practices but also develops it. One important development is to do with the way in which family practices are implicated in 'boundary work' (McKie & Cunningham-Burley, 2005). Some of these boundaries (both spatial and temporal) are externally imposed through, say, licensing laws or health and safety legislation. But these boundaries overlap with or are coterminous with the more fluid boundaries constructed on a day-to-day basis through preserving distinctions between home and work, between back-stage and frontstage. Thus, this particular study illuminates some of the spatial and temporal dimensions of family practices (see Chapter 5). Another way in which this study develops the core ideas of family practices is in the way in which it illuminates the relationships between 'families we live by' and 'families we live with' (Gillis, 1996) that is, roughly, between discourses and practices. This distinction is normally assumed and somewhat subliminal in everyday life. However, there are occasions such as family crises (the experience of disability, for example) where this distinction becomes apparent. It can be argued that these somewhat atypical arrangements where the home is also a work-place also provide occasions for the exploration of these differences as family members seek to construct domestic life and relationships in an environment which, to an external gaze, might seem much removed from the expectations of family life.

Displaying families

An example of an article which does not simply refer to or use the ideas of 'family practices' is Janet Finch's discussion of 'displaying families' (Finch 2007). This article, which itself has generated a considerable amount of interest within the UK, takes the approach of 'family practices' as a point of departure. In particular, the ideas derived from the practices approach are those of 'doing' family and the overall emphasis on fluidity in the definition of family boundaries.

The core argument is 'that families need to be "displayed" as well as "done"' (Finch, 2007: p. 66). The concern, therefore, is with conveying meanings of family to significant others:

> Display is the process by which individuals, and groups of individuals, convey to each other and to relevant audiences that certain of their actions do constitute 'doing family things' and thereby confirm that these relationships are 'family' relationships.
>
> (Finch, 2007: p. 67)

Finch provides several reasons why the idea of display is not only an interesting extension of the idea of family practices but might also be seen as significant in the context of the changing meanings and family practices in the UK today. The first is the increasing realisation that family is not to be equated with household. While this has always been an important conceptual distinction, a variety of changes have meant that this may be of greater significance today. For a variety of reasons – divorce, separation and reconstituted families and the growth of transnational kinship ties for example – there is an increasing need to look at relationships between households rather than to assume that family relationships are chiefly confined within them. Associated with this, one might add, is an increasing fluidity in the idea of 'home'.

Second, and linked to this, is an increasing fluidity of families over time. In other words there is a decreasing sense of a more or less standard life-course as individuals move in and out of different residences and different sets of relationships and as the sequencing of these transitions becomes less and less predictable. What these changes mean is that the work of demonstrating the range and character of one's family relationships becomes more complex and requires more 'work' on the part of the relevant participants.

Finally there is the changing relationship between family and personal identities. Referring to the influential writings of Giddens (1992) and Beck and Beck-Gernsheim (1995; 2002), Finch points to an increasing emphasis on choice within intimate relationships, some of which may be defined in family terms while, for others, it may be important to mark out a clear distinction between family and other chosen relationships such as friends or same-sex couples. The blurring of the boundaries between kinship and friendship relationships constitutes another part of the background for a contemporary study of family display and a possible reason why this has become an important area for consideration in the early twenty-first century.

Display, in Finch's terms, is a way of stating 'These are my family relationships and they work' (Finch, 2007: p. 73). The need for stating this, where such a need is felt, will clearly vary between individuals and their circumstances. Points of ambiguity or unscheduled transitions might provide such a context where the need to display family might be felt more acutely than on other occasions. Another illustration might be provided by the already mentioned article by Seymour (2007) where family members occupy the same living area as hotel guests or non-related employees of the establishment. Other occasions might include where family members place themselves, potentially at least, within the public gaze such as meals in a restaurant or events such as degree ceremonies.

Finch also refers to various 'tools' for display. These include material objects that are given significance by being enmeshed within webs of meaning – heirlooms and keepsakes – or family narratives or jokes. Another example provided by Finch is a later article to be found in naming practices (Finch, 2008).

From a global perspective, naming practices vary considerably and deal with matters of considerable complexity. However, even the model with which most readers of the article will be familiar, forename plus surname, raises a whole range of important, and relatively unexplored issues. Making an important distinction, Finch states that 'a name is both a legal *identifier* of the individual but also a potential part of the social *identity*' (Finch, 2008: p. 712, emphasis in original). Naming raises both issues of individuality and connectedness.

Finch links this discussion of naming to her earlier discussion of family display. Many conventional naming practices – the routine assignment of the father's surname for example – are taken for granted to the extent that they are rarely seen as illustrations of family connectedness or family display. The choice of a forename is formally seen as being more open but, in fact, frequently involves the use of the name of a relative and may hence also be a form of family display. Where these issues become important, as with the wider issues of display in general, is where conventional patterns do not necessarily provide clear guidelines as, for example, in the cases of adoption or the naming of children with lesbian parents.

To return to the theme of displaying families, it can be seen that this a fruitful extension of the idea of family practices and one which is itself, as Finch recognises, capable of further elaboration. Here, I suggest three possible extensions which might be of importance:

- It might be useful to explore the overlap between class practices and family practices using the idea of display. An obvious example might be the display of members of the Royal Family on the balcony of Buckingham palace. It is possible that such displays were more important in the past but we might also consider figures in the public eye such as politicians and celebrities.
- There is plenty of scope for the exploration of the different kinds of audiences involved in the display. Some might be quite distinct from the family members in terms of actual connection while others might be linked in various ways to the set of people engaged in the display. Some forms of display, therefore, might be like a game of charades where, at different times, those displaying are also in the audience.
- It might be interesting to explore the difference between 'displaying families' (the core of the original argument) and 'displaying Family', that is the deployment of family members in displaying the idea of and the core values attached to family. Notions of 'Christian family living' might have something of this character.

Critical issues

To a very large extent, references to the idea of family practices have been largely favourable or, at least neutral in tone. Scholars have either simply referred to the term with little elaboration or they have developed or used the concept, highlighting one or more features that have been found to be particularly useful. With two or three exceptions, more negative comments have been rare. I shall refer to these more critical comments in this section but I have also found it necessary to reflect more critically upon my own ideas and possible limitations. This section is, therefore, to some extent a discussion with myself.

Use of the word 'family'

I have argued that one of my main aims in developing the idea of family practices was to move away from discourses around '*the* family' with all the reificatory potentiality that this might suggest. However, it might still be argued that my talk of *family* practices (as opposed to some other kind of practices) tends to perpetuate a particular set of understandings which, however unwittingly, place the family at the centre of the analysis. It could be argued that some of the alternatives that were suggested in the previous chapter – intimacy, personal life, the 'total social organisation of labour' and 'caringscapes' for examples – implicitly suggested that my focus was too narrow. By writing about 'family practices' as opposed to, say, intimate or personal practices, I was still placing the emphasis on family relationships to the exclusion of other kinds of relationships which, it might be argued, were becoming increasingly important and which were certainly more inclusive. A focus on family practices could easily slide back into business as usual as far as family studies are concerned.

I think that it is true that my analysis does focus upon families as conventionally understood (relations between spouses, parents and children and between kin) with relatively little reference to other kinds of relationships. This is certainly true for most of my illustrative material and it is probably true for the general drift of the argument as a whole. I shall defer a fuller answer to the question about the distinctive or special character of family relationships in my concluding chapter but here I shall just mention three reasons why I continue to use the term 'family practices':

(a) As I argued in the previous chapters while all these proposed alternatives all add important dimensions to the analysis of family and other relationships they tend to overlap with rather than to be coterminous with family relationships. In many cases these alternatives may be broader in scope than family practices but they are not the same and they raise different problems of analysis and interpretation. Thus, depending on your definition of 'intimate', not all family relationships are intimate. Similarly, not all family relationships are caring relationships and so on. I see this whole range of approaches (including family practices) as collectively providing a very powerful set of tools that can be used according to the problem under investigation.

(b) 'Family' still appears to be an important matter of concern to large sections of the population. While notions of friendship are clearly also seen as important, perhaps increasingly so, there is little evidence to suggest that they are being perceived as replacing family relationships however these relationships may themselves be changing in scope and complexity.

(c) 'Family' still occupies an important section in scholarly and public discourse over several scholarly disciplines. In some representations 'intimate relations' or 'friendship' may be added to family but they are not seen as alternatives. The extension of the idea of 'family practices' into 'displaying families' still highlights family relationships while giving clear recognition to other kinds of intimate or personal relationships.

Thus, for the time being at least, I shall still write about 'family practices' while recognising that the very fluidity of the term as I have developed it allows me to think about intimate or personal practices as well. Beyond this, there is a reminder to myself that I should continue to use the word 'family' with some care even when it is linked to 'practices'.

The question of structural constraints

In writing about a whole range of recent approaches to the family, including 'family practices' as well as 'chosen families', Widmer et al. make the following critical comment:

> These concepts, however inclusive they may be, do not convey the fact that family relationships, because of their complexity, are likely to remain highly patterned and embedded in the social structures of late modernity. Emotional, financial, educational, and domestic resources are scarce, and individuals do not decide about their allocation in a social vacuum where only their self-interest or some culturally predefined 'life style' rule.
>
> (Widmer et al., 2008: 3)

This comment is made in the context of developing an argument about 'family configurations' which I discussed in the previous chapter. Here I should reject the references to 'self interest' or 'life style' as being applicable to my analysis. However, I am also willing to acknowledge some degree of truth in the argument as a whole. My approach to family practices, at least in some respects, does seem to overemphasise agency at the expense of structure; at least the references to 'doing' family and fluidity would seem to suggest as much. The question is whether it is possible to recognise the strength of this critique while continuing to develop the family practices approach.

Widmer et al. refer to two sets of social constraints which may limit the 'doing' of family in the ways suggested by my analysis. The first refers to the patterned and deep continuities within family relationships which continue over time and which remain 'embedded in the social structures of late modernity'. Family configurations, while there may be some element of choice within them, do not lose their ascribed character. Even those relatively few individuals who cut themselves off from all known family ties (or those more numerous individuals who are cut off as a result of war or natural disasters) will have some influential memories of past family relationships.

The second set of constraints refers to those resources which are available to family members, which are allocated within families and households and which, almost inevitably are scarce. Some of these resources will be part of the family inheritance, the social, cultural and economic capital, while others will become available to family members through their engagement in employment, education or other spheres of life. The amount, the character and the allocation of these resources is not something which can be determined by any one family member. Together, the delineation of these two sets of constraints provides strong reinforcement of the argument that family practices are not conducted in a vacuum.

It can be argued that these sets of constraints impact upon family practices, the 'doing' of family, in two rather distinct ways. In the first place, the fact that people may 'do' family or carry out family practices does not necessarily mean that they willingly chose to do so. While notions of the reality or otherwise of any choices are always problematic, the linking of practices to 'habitus' may remind us that practices may be performed in a routine, taken-for-granted manner. This sense of relatively unquestioned routine has been part of the family practices package although perhaps it has not been emphasised enough up to now.

An example of this kind of taken-for-granted family practice may be taken from the extensive feminist scholarship around the theme of housework or domestic labour. For many women these activities seemed to be given as a consequence of becoming a wife and a mother (identities which, themselves, might have been chosen in only a very loose sense of the word). They were practices in that they were orientated to other members of the family and that they reproduced the very idea of family and the identities subsumed within it. But they were not chosen or, necessarily, enjoyed. Something similar might also be argued around the bundle of expectations and responsibilities that were associated with being a 'breadwinner'. For both mothers and fathers family practices were merged with gendered practices and it could be argued that this merging contributed to the sense of inevitability and necessity associated with these practices. In this sense the family practices were not only shaped by constraints but they also constituted these constraints and their reproduction over time.

There is another way in which family practices may be associated with constraints and this is probably closer to the argument of Widmer et al. Individuals might wish to 'do' family in a particular way, to be 'good' parents, 'good' partners and so on but feel constrained, through the scarcity of key resources, from doing so to the fullest extent. It may be argued that these notions of 'good' themselves constitute another kind of ideological constraint but nevertheless they do highlight a disjuncture between what is desired and what is practically achievable. Studies of low-income households and lone parents as well as of families confronting long-term illness or disability provide detailed explorations of these disjunctures. Yet again, issues of the use of time as a resource within families and questions of work-family balance may show the kinds of constraints that confront family practices on a day-to-day basis.

Thus family members may feel constrained through the routine habitus of family practices and they may feel constrained from doing family in a manner that they would ideally wish. It can be argued that my original formulation gave scant recognition to these constraints. Nevertheless to recognise these constraints does not invalidate the family practices approach but may, indeed, provide a way of deepening and enriching it.

Questions of Discourse

A possible criticism of the family practices approach, although not one that I have actually seen, is that it pays too much attention to practices

at the expense of discourses. It is certainly true that my original statement had little, if any, attention to questions of discourse. To use
another influential distinction, my account concentrated on 'the families we live with' rather than 'the families we live by'(Gillis, 1996).

Why does this matter? One answer might be to argue that discourse
constitutes the context of constraints within which family practices
are conducted. If we take the stronger notion of Discourse (the one
identified with Foucault) then it represents a particular combination of
knowledge and power. Thus notions of 'dysfunctional' families may be
elaborated (through mixtures of professional and political discourses) as
part of a wider understanding of a 'broken society'. Such models, which
also implicitly play on ideas of 'functional' families, may shape what
may be said or not said about contemporary family life substituting, say,
notions of family or individual pathology for discussions of structured
inequalities.

To what extent do these wider discourses about families, functional
or dysfunctional, actually impact on routine family practices? This is,
clearly, a complex matter and we no longer (if we ever did) see family
members as being passively at the receiving end of discourse or, to use
an older term, ideology. Individuals can and do negotiate with discourse
partly because the public accounts of family are rarely completely
coherent or uniform. There are frequently competing versions of what
the family is about which are publicly available. Nevertheless, individual family members are clearly aware of wider family discourses and
these may influence, say, step-parents or lone parents as they go about
their routine practices even when, as may frequently be the case, they
deny the applicability of these public accounts for their daily lives.

What becomes clear in following through these, or similar, examples, is that practices and discourses, the families we live with and the
families we live by, are mutually implicated in each other. Consider the
pairs of terms, mothering and motherhood, fathering and fatherhood
and parenting and parenthood. Broadly speaking, the first term in each
pair represents practices while the second represents discourses. It might
also be noticed that the first term is, in English, the newer term. Its
equivalent does not exist in some other languages and even in English
we have not yet got around to providing a companion term for 'childhood'. This might suggest that there has been a long-term shift away
from the public statuses of motherhood, fatherhood or parenthood to
a more active understanding of what mothers, fathers or parents actually do. But this does not necessarily mean that discourse is absent.
In engaging in activities associated with being a mother, a father or a

parent (and whether the gendered or gender-neutral term is deployed is itself a matter of some significance) family actors are looking in two directions. The one is to the business at hand, the everyday and possibly routine family practices while the other is towards more public discursive constructions of mothers, fathers or parents. The desire to be a 'proper' mother or a 'new' father will enter into these daily practices, even where there may be some degree of distancing from these public identities. Further, discourses are not produced in a vacuum. Discourses themselves draw upon practices. A good example of this circularity is to be provided by the 'Family' supplement of the British newspaper *The Guardian* published on Saturdays. These contain two or three feature articles about how individuals have responded to complex family problems (parents who belong to an all-embracing religious organisation, a father who saw himself as a literary failure, for example) together with some shorter regular items in which readers are invited to contribute and discuss the importance of particular family photographs, favourite foods and music which has some particular family significance. The supplement and similar publications draw upon everyday practices as well as more extraordinary events in order to construct a complex, multi-sided discursive picture of family life in general. In these accounts the everyday is given some significance and the dramatic is also incorporated into a wider understanding of what 'the family is really about'. (It is interesting to note that this supplement is clearly separate from other supplements which deal with 'Work', 'Money', 'Leisure'. Etc.)

I conclude, therefore, that a full account of family practices must also include references to these more discursive considerations if only because the two are so intimately related. Yet even these brief illustrations should indicate the complexities of the inter-connections and the need to explore these in more detail through the working of particular examples.

Family practices and social divisions

One complaint which might be made about much family sociology, including the family practices approach, is that it implicitly represents the experiences and world view of a relatively small section of global society. This section might be characterised as Anglophone, white, middle class and probably male. Issues of sexuality, where this critique has been made explicitly, I shall defer to the next subsection.

I think that this potential objection as far as gender is concerned can be dealt with quite easily and reflects my argument that while most, if not all, human practices are gendered there are very few practices which are simply about gender and nothing else. In the case of family practices I have argued that one of the characteristics of the practices approach is the sense of fluidity and that what might be understood, from one perspective, to be family practices might also be seen, from another angle, as gender or some other practices. The fact that the overlap between family practices and gender practices is often closer and more frequent that the overlaps with other practices needs to be taken into account but does not in itself invalidate the general argument. For the most part it could be argued that family practices can also be seen as 'gendered family practices'. At all times I, and others who want to develop the idea of family practices, should always be sensitive to the strong overlap between gender and family practices.

Much of the same argument can be adapted to considerations of class and ethnicity. In terms of the illustrations that I provide, I am fairly certain that a white, middle-class bias can be detected in my work. Simply to recognise this as a potentiality does not deal adequately with this accusation. My only hope is that others with different personal and intellectual biographies will still find the overall approach useful and will seek to adapt it to different bodies of work. One of the motives for developing the practices approach was to develop an orientation that was not based on a particular normative and limited model. It is clearly important and possible, within this approach, to see the ways in which family practices overlap with class and ethnic practices and to explore how these practices serve to reinforce or modify these other social divisions. There are, in fact, several studies which do this (e.g. Becher, 2008; Jordan, Redley & James, 1994; Savage et al., 1992) even if there may or may not be direct reference to family practices.

The more difficult issue relates to the wider accusation of ethnocentricism. To what extent is the very idea of 'family', and hence family practices, based upon a particular set of experiences located in what might be variously defined as 'The West', 'Anglo-phonic countries', 'The First world' and so on? This argument is a familiar one and one certainly not limited to discussions of family. Indeed, there is an argument to the effect that many of the key terms deployed within social science are, in fact, limited in this way (Bhambra, 2008).

These are not issues that can be dealt with quickly or lightly. Indeed, they deserve another book. However, part of my response would be to recall that one strand in my approach to family practices was to see it in terms of the interplay between the researcher and the researched, between the outside analyst and the everyday activities of those being analysed. The term 'family practices' is an abstraction of the kind to be found in any other scholarly discussions of family life. The point is whether the use of the term can both get us closer to the ways in which individuals experience and understand their family lives and/or whether it provides new ways of understanding these experiences from the outside in, for example, suggesting new connections with other areas of social life. The test, ultimately, is a pragmatic one.

Sexualities

One specific criticism of the family practices approach is that it continues (along with other modes of family analysis) to present a heteronormative model of intimate relationships (Roseneil, 2005). The approach tends to underplay the degree of social change that has taken place within the field of intimate or personal relationships and, in particular, undervalues the importance of friendship in modern society:

> Postmodern living arrangements are diverse, fluid and unresolved, constantly chosen and rechosen and heterorelations are no longer as hegemonic as once they were.
>
> (Roseneil, 2005: p. 247)

I think that there is a certain degree of truth in the claim that my discussion does tend to present a broadly heteronormative view overall. This is almost certainly true for most of the illustrations that, at least up to now, I have chosen. And the use of the word 'family' to limit the term 'practices' has, without any further qualification, a strong presumption of heterosexuality. Insofar as this is not challenged with sufficient vigour there must be some justice in the change of my perpetuating heteronormativity.

But is this inevitable, something built into the very idea of family practices? Weeks, Heaphy and Donovan, in the already referred-to study of same-sex intimacies, draw upon the idea of family practices with some approval (Weeks, Heaphy & Donovan, 2001: p. 37). They draw, in particular, on the active emphasis and on the move away from a more institutional-based understanding that is implied in the practices approach. The practices they describe are activities that many of the people within their sample would recognise as being of importance:

> For ... them, 'family' is about particular sorts of relational interactions rather than simply private activities in a privileged sphere.
>
> (Ibid.)

Of course there is a continuing debate about whether the language of 'family' is appropriate in accounting for same-sex couples (or, indeed, any other 'life experiments', heterosexual or otherwise). Doubtless some of the individuals in this, and similar, studies, would clearly be highly suspicious of any use of the term 'family' and would provide strong arguments in favour of this rejection. Nevertheless the fact that at least some individuals and couples within single-sex relationships may chose the word 'family', however qualified, to describe their practices must be a matter of some significance.

There are two further points to be made. The one is that, whatever an individual's chosen living arrangements and however central these are in that individual's core life projects, there will nearly always be some family relationships in that individual's past and which may still have an influence. Even a chosen rejection of these early ties in their entirety may, negatively, attest to the significance of these ties. These ties may form a numerically small proportion of the sum total of personal ties but they are still there.

And this leads me to the second point. I have written about family practices because I find family processes to be of interest, intellectually, and because I also recognise that they continue to be significant, emotionally and practically, for large numbers of people. It may be possible to extend the practices approach to a whole host of non-familial intimate or significant relationships; indeed, I believe this to be the case. However, as with any field of enquiry I must recognise the limitations of my field. All I can attempt to do is to try to ensure that my use of the term 'family' does not entail signing up to any one normative model of family living.

Over-optimistic?

One final possible criticism of my approach, although not one that I have actually encountered, is that it presents an over-optimistic perspective on family life. This, again, may reflect the range of illustrations that I chose which may highlight a relatively positive, benevolent picture of everyday family living. This, of course, can be rectified, if not by me than by somebody else wishing to develop this approach. It might also be implied in the emphasis on 'doing' which has a positive ring to it, more than a hint, perhaps, of 'yes, we can!'

If this be the case, it may reflect something in my own personality or it may reflect a desire to avoid a 'problems' approach to family living. A popular view of sociological enquiry is that it is concerned with issues defined as social problems, in this case with problems of divorce, lone-parenthood or abuse of children, rather than with more prosaic features of everyday living. In contrast, I have preferred to explore sociological rather than social problems, that is to explore matters which I find puzzling or interesting rather than disturbing or demanding intervention.

However, this is probably a less than adequate response and, indeed, in earlier writings I have criticised some approaches to family studies as ignoring the darker side, especially the violence and the abuse (Morgan, 1975). There is no doubt that this dark side still exists but there is no reason why the family practices approach should not give recognition to this. Family practices may well be cruel and abusive, whether intentionally or unintentionally. To call such practices family practices is not to invest them with an aura of virtue but simply to say that that they are carried out with reference to others we are defined as being family members. Further, ideas of the family, enacted on a day-to-day basis as practices, may perpetuate a limited, even selfish or destructive perspective on the world. Practices conducted 'in the name of the family' may have implications for the perpetuation of social inequalities or for the destruction of the environment. I need to recognise this and to incorporate this in my analysis but there is no logical reason why a practices approach cannot confront these darker questions.

Conclusion

In this chapter I have aimed to show that the practices approach has had some kind of resonance within the sociological community (especially that subsection concerned with family life) and that it has been referenced and used in a variety of ways. In some cases, as with displaying families, it has been developed in such a way as to raise new issues and areas for research. All this attention has been gratifying.

Gratifying, too, is the fact that there have been relatively few negatively critical comments on my work although perhaps there should have been more. I have attempted, here, to deal with these and with some other possible criticism that might be levelled at this approach. Doubtless there are more. However, my overall conclusion is that the family practices approach still has potentiality for further development and that many of these criticisms, real or possible, can be incorporated into these developments. I shall try to show this in the chapters that follow.

5
Time, Space and Family Practices

Introduction

Issues of time and space are necessarily implicated in family practices (Daly, 1996). As an illustration consider a child in a reconstituted or bi-nuclear family, that is where he or she moves between the homes of the two, now separated, original parents (Haugen, 2007). In this not unfamiliar example there is a particular time (weekends, holidays) which is the subject of negotiation, sometimes with legal interventions, between the two parents and their new partners. Simultaneously, this involves at least two spaces and the movement between them. It is difficult to think of any family practices where considerations such as these, dealing with time and space, do not arise.

Further, as has often been noted, time and space are closely linked to each other. 'When?' and 'Where?' are two key questions that are raised in relation to any social arrangements; '2.30 outside the supermarket'. Urry's discussion of 'co-presence' with its three related themes of 'face to face', 'face the place' and 'face the moment', brings this out very well (Mason, 2004; Urry, 2002). However, while there is mutual implication there need not be complete overlap. Birthdays are points in time while a birthday celebration exists in time and space. Remembering a birthday may involve disjunctures in time and space.

The way in which time and space are involved in everyday family practices and with each other can be seen in the everyday phrase, often the subject of ironic comment, about 'spending time with the family'. This is also, inevitably, 'sharing space with the family'. 'Quality Time' would be lacking in quality if it did not imply physical co-presence as well. 'Me time', a term denoting time spent in reference to but away from other family members, might also involve a separate space apart from these others. When grandparents talk about their role of 'being there' for their grandchildren or of 'waiting in the wings' they are clearly referring to both spatial and temporal considerations (Mason, May & Clarke, 2007).

Variations within space

Although time and space are mutually implicated in each other, each raises slightly different questions in relation to family practices. It may be useful, therefore, to begin by considering these two sets of themes separately. In the case of space there are a cluster of issues around ideas of home, domestic space and distinctions between the public and the private. In writing about home, in the context of examining young people leaving home, Holdsworth and myself found it useful to adapt some ideas from Lefebvre (Holdsworth & Morgan, 2005; Lefebvre, 1991). Here we explored distinctions between 'practical', 'symbolic' and 'imaginary' space:

> In referring to 'the practical' we are concentrating upon the more material, in the broadest sense of the word, aspects of domestic space ... With the 'symbolic' we are referring to discourses about and representations of home and domestic space. For the 'imaginary' we unpack how individual meanings are inscribed into domestic space.
> (Holdsworth & Morgan, 2005: p. 75)

Looking at family practices more generally, the practical refers to the ways in which space is organised in order to achieve practical ends. In one sense this can be seen as being at the heart of what is implied in family practices. These include divisions of labour within the home and between the home and outside, divisions which may well map on to differences between genders and generations. These spatial practices are not simply internal to the home they also involve outside relations involving the triangle of family, state and market. Decisions about where to live, for example, may involve not simply the design and the cost of the accommodation but also the ease of travelling to and from work and the location of amenities such as schools and medical facilities.

In terms of the symbolic we are looking at all the meanings attached to and discourses concerning the home and family practices in general. These include notions as to the centrality of family life, ideas of a 'proper' or a 'real' family and of the linkages between ideas of home and family and notions of individuality and privacy. These numerous ways in which family life is represented and evaluated do not necessarily form a coherent whole; indeed, in a complex late-modern society, there may well be contradictions and negotiations between different conceptions of family.

The imaginary deals more with individual ideals of family life and of the settings of family practices within the home and beyond. Imaginary family practices may reflect individual biographies and the ways in which these are shared and merged in the course of the conduct of family practices. We are looking at the way in which the locations of family practices, chiefly but not exclusively the home, reach beyond in the imaginations of those participants to dreams, desires and hopes. Thus, a child's bedroom may reflect imagined notions of childhood and themes of security and the imagination.

All three of these sets of themes, the practical, the symbolic and the imaginary, interact with each other throughout the performance of family practices in the various family-related sites. Richards' discussion of the development of an Australian middle-class suburb illustrates these interconnections very well. Practical considerations to do with mortgages, building and decorating interact with the symbolic meanings attached to the home and its location as well as with more individual dreams and aspirations (Richards, 1990). More generally we can see these interconnections in everyday processes of 'home-building' and 'home-making' and DIY. Again there may not always be congruence in terms of the perspectives of the different family members. For different individuals and at different times the home may be a haven or it may be a prison or often a complex mixture of the two.

Again we can see the connections with time. The imaginary in particular might be shaped out of constructed pasts and imagined or hoped-for futures. Locations of home may revolve around notions of the working day, week or year and the movement between different sites. Individuals have a variety of housing careers (Pickvance & Pickvance, 1995) and these movements too enter into the practical, the symbolic and the imaginary constructions of family spaces.

Urry's discussion of co-presence also touches on similar variations within the idea of space:

> One should investigate not only physical and immediate presence, but also the socialities involved in occasional co-presence, imagined co-presence and virtual co-presence.
>
> (Urry 2002: p. 256)

This is argued in the context of a more general discussion of mobilities but it also has relevance when we are considering family practices and serves as a reminder that the performance of such practices does not always require the physical co-presence of the other.

We should also remember that the spatial dimension of family practices is not confined to the place identified as home. A parent's place of work, a familiar location for weekend outings, regular holidays with relatives; all these provide examples of the way in which family practices extend beyond the home, indeed sometimes across continents. In the case of migration or major disruptions due to war or natural disasters the links between family practices and home (expect perhaps an imagined home) may be severed altogether. Similarly, the place identified as home need not be limited to family practices. Homes exist in localities and neighbourhoods and may be the base for extensions into widening sets of friends and acquaintances.

Variations within time

Sociological and anthropological discussions of time contain numerous distinctions to provide a basis for analysing its significance and impact in everyday life. Southerton (following Fine), for example, refers to five dimensions: Duration, Tempo, Sequence, Synchronization and Periodicity (Southerton 2006). Using distinctions such as these, he is able to explore such topics as the sense of harriedness experienced within many families. The experience of time being squeezed was common among his sample of twenty households (Southerton, 2003). They referred to 'hot spots' where numerous predictable time constraints converged (combining getting to school and getting to work in the morning) and the less predictable but highly desired 'cold spots' where individuals could relax and enjoy 'me time'. The ways in which these intensities of experienced time might be exacerbated is well described in a study of mothers with children with attention deficit hyperactivity disorder (Taylor, Houghton & Durkin, 2008). There is a constant problem of dealing with early rising or reluctantly rising children and the impact of these daily struggles on other family members.

Further distinctions might be made in parallel with the discussions in the previous section. Although these distinctions between the practical, the symbolic and the imaginary were developed in relation to space, domestic space in this case, similar considerations arise in relation to time. Thus we may talk about practical time, referring to the everyday routines, the habitus. Many family practices are of this kind involving the same or similar tasks being repeated with clearly detectable regularities. Practice, in the sense of repetition, may not make perfect but it does make practices.

These routine day-to-day or weekly practices also merge with longer-term regularities which have a more seasonal character. These may be regular events (birthdays, for example) which may be celebrated within a family context or national or religious events which, again, usually include ideas of the family as a base or a core unit. The strong associations between ideas of the family and the celebration of Christmas for example are well known and here again we have a coming together of the practical (the numerous preparations of food, decorations and gifts), the symbolic (the discourses that link the sacred and the secular) and the imaginary.

The practical time associated with daily routines and seasonalities may contrast with a different sense of time, namely that associated with the special, the exceptional or the memorable. These may be family-based events associated with births, deaths and marriages or anniversaries. Or they may represent special achievements such as graduation. Such special events may not always be positive or conflict free. The funeral of a child or a young person may be an obvious example but so too may be divorces and separations, family crises associated with accidents or ill-health or simply occasions, such as holidays which 'went wrong'. All these exceptional events, as well as the more routine and the seasonal, involve practical considerations and the necessary allocations of time, individual or shared, to particular activities.

A good example of the way in which family practices routinely link different understandings and meanings of time is provided by Becher's study of South Asian Muslim families in Britain (Becher, 2008). She writes about the overlaps, and occasional tensions, between family time and sacred time; ' sacred activity was viewed as an integral part of domestic and family life' (ibid.: 88). Ramadan, for example, had many strands of meaning. It was seen as a special time for the practice of religion, as an opportunity to link to kin and community and defined a sense of the wider Islamic world. It also could be linked to the growing maturity of children as they were seen to be 'old enough' to participate fully in the periods of fasting. Thus family and sacred time were also linked to the life-course.

It has already proved to be difficult to talk about practical time and family practices without also bringing in the symbolic and the imaginary. In the case of the former we are talking about either the positive values attached to time spent on family tasks or the sense of duty and obligation which is expressed as a way of accounting for these particular expenditures of time. The notion of 'putting the family first' contains some clear temporal connotations.

It is worth concentrating a little further on the imaginary aspects of time and family practices as these tend to be rather under-explored as compared with the more practical or symbolic aspects. This involves

The Past (memories, nostalgia)
The Future (projects, hopes)
The Present (the meeting of pasts and futures)

While it should be possible to apply this set of descriptions to almost any individual in any social situation, it is likely that it has a particular resonance in thinking about family practices. This is because not only do family members use and negotiate about time on a regular basis but they are also located within time in certain distinct ways. Indeed, in a certain sense it could be argued that family practices are *about* time. This arises out of certain features of the life-course, seeing this in relational rather than in individual terms. The co-presence of family members is a co-presence over long and overlapping periods of time such that the life-course of any one individual is lived out and through in necessary relationship with the life-courses of other family members. Thus, pasts and futures are rehearsed and recollected in the co-presence of these others, of different generations.

As an illustration, consider two sisters sorting through their mother's possessions after her funeral. Among these possessions are some photograph albums and a considerable amount of time is spent in going through these, trying to identify barely remembered relationships, sharply remembered other family gatherings and past joys and sorrows. Links are made across generations, not only in the past but perhaps also in the future when decisions are made about what is to be passed on to younger members of the family.

There are two important features of this discussion which require stressing. The one is that it is frequently not only individual or family time that is being explored in these temporal practices. It is also historical time and here we see the links between history and biography which we see as a key element in family practices. In remembering a recently departed family member we may recall that person's age at the time of death or we may state the individual's date of birth. They both give the same information, but the latter conveys a sense of historical generations and past historical events (World Wars, the Great Depression) that impacted on these past lives. It is often through conversations, real or imagined, between different family generations that we gain imaginative access to a historical past.

The second is to emphasis that the relationship between time and family practices is based upon relationships or family configurations and not simply upon individuals. Family stories and family myths are elaborated not by individuals and rarely by clearly bounded family groups but by different and overlapping smaller sets of relationships that themselves change over time. This complexity and richness in part explains the popularity of the exploration of family histories and the popularity of programmes such as the BBC's *Who Do You Think You Are?* In which a well-known person traces part of her or his family tree.

Variations within time and space

I have, briefly, considered issues of space and time in relation to family practices separately because each raises slightly different problems. However, it will have become apparent that it is very difficult to separate these and for the rest of this chapter I shall take them together.

Both time and space, in relation to family practices can be seen either as constraints or as resources. Within any one family configuration both may be relevant. We recognise the constraints when individuals talk about 'not having enough time' or of feeling overcrowded. The pokiness of a room or the pressures of time are, of course, not absolutes but are to be seen in relation to family projects. At the same time family practices involve the use of, the negotiations about, time and space. Sequencing of activities – you can watch the DVD after you have finished your homework' – is a familiar feature of everyday family practices. Similarly family custom and practice may mean that not all family members have equal access to all parts of (or objects within) a family home. Put another way, family practices are conducted *within* time and space and involve the *use* of time and space.

Timetables (which may have an actual physical manifestation or which may exist at a more tacitly understood level) are clearly examples of time/space as constraints and resources. (Although we use the word '*time*tables' the spatial aspect is implicitly recognised.) These may be constructed in relation to external constraints (the organisation of paid employment, school and opening hours, public transport timetables and so on) but are not wholly determined by these. Within this framework, decisions are made about who does what and when. The basis of established priorities may be expressed in terms of what may be described as 'legitimate excuses' (Finch & Mason, 1993). Whether an excuse, such as a claim that one is a bit pushed for time at the moment, is heard as a legitimate one is part of the everyday negotiation within family practices. Decisions, arising out of the negotiation of timetables, may be routinised as part of everyday family practices or the family habitus, part of the 'nomos creating' aspect of everyday family life (Berger & Kellner, 1964). These timetables will themselves be subject to change over the life-course, as children leave home or adults retire.

There are two sets of distinctions which can be made in relation to time/space and family practices. The first is that the practices can be described as being either strongly bounded or weakly bounded. Note that the reference here is to 'family practices' and not to types of families. We may also refer to families in this way but our concern here is with the practices. The same 'family' might engage in both strongly bounded or weakly bounded family practices.

Strongly bounded family practices include strong and more explicit references to 'our' family and clearer distinctions between 'them' and 'us' expressed in family terms. The contrast might be between 'close' family and other family members, between family and non-family or between a particular configuration and outside agencies. An example here might refer to funeral arrangements where there may be a contrast between a service for close family members (or private mourning) and some more public or open celebration of or a memorial to the deceased.

Weakly bounded family practices, in contrast are to be found where there appear to be relatively few boundaries between family and non-family or where non-related persons might be treated, under particular circumstances, as 'part of the family'. There may be relatively easy access to a greater number of 'family spaces' or the sharing of time. An example might be found where the home is regularly used as a basis for voluntary activities such as working for charities or political or religious organisations. Michael Rosen talks about growing up in a Jewish Communist Party family in these terms:

> Branch meetings used to take place in our house so I have got a very early memory of groups of people turning up and coming round the house. I have a sense that my parents had a circle of friends and these were party friends.
>
> Because my parents knocked around with other people in the party who had families, there was a culture of 'CP Kids' who would then

talk about the party and Russia and things like that. We would meet on social occasions.

(Rosen/Bloomfield, 1997: pp. 52–3)

There is a related, but distinct, set of differences to do with the relationships between time/space and family practices. This is a distinction between concentrated and diffused family practices. Concentrated family practices are those where there are close links between different sets of practices, a number of different activities which are clearly related (spatially or sequentially) to each other. To a large sense these practices are mutually reinforcing and provide for a greater sense of overall coherence. Examples of these are provided by family ceremonies or family crises which may involve sudden, if temporary, shifts of location and changes of timetables.

Diffuse family practices are those practices which are carried out individually or with small, possibly short-lived, configurations of family members. There may be little reference to other sets of people or practices. There is little requirement for the co-presence or even the co-awareness of other family members. Travelling home from work might be an example. It is an individual and work-related practice and may, in this somewhat liminal time and space, have its own immediacy and interest. But the fact that the individual is travelling home (and not somewhere else) may make it a family practice. Similarly, the fact that the individual is making this particular journey at all may be a consequence of earlier family practices to do with decisions about where to live. Another example of diffuse family practices might be the numerous family errands which individuals might conduct in the course of a week. A trip to the shops may be made with other family members in mind (their tastes, their particular timetables) but it is frequently carried out by individuals and the trip itself may take on its own, if temporary, reality. But it is still a family practice if not necessarily exclusively so.

Most discussions of family practices (whether the word is used or not) implicitly tend to deal with practices which are assumed to be both strongly bounded and relatively concentrated. The worlds created by family sitcoms are worlds where a relatively small number of related people do things within relatively limited periods of time and space and which are clearly linked to each other. This, of course, has as much to do with narrative conventions and the constraints of plot. Yet some actual families might seem like arenas of perpetual busyness and careful, if hidden, orchestrations. This is partly because of the close overlaps between, and sometimes confusions between, family, home and household. But, in contrast to this, within late modernity, it becomes increasingly necessary to consider family practices which are often more weakly bounded and more diffuse.

My first illustration of these discussions is one that has already been mentioned; the positions and experiences of children following the divorce and re-marriage (or re-partnering) of their parents (Haugen, 2007; Smart & Neale, 1999). These, increasingly familiar, experiences produce what has sometimes been called 'bi-nuclear' families pointing to the way in which two separate households have been established through this process, with the children moving in between.

There are several general points that may be made about these experiences. The first is that they entail a shift from relatively taken-for-granted routines and family practices, a family habitus, to something which is much more subject to deliberation and negotiation. Tacit rules have to be redefined and made more explicit. Much of this involves a reassessing and a redrawing of spatial and temporal boundaries. From the point of view of the child this may entail experiencing two domestic environments, their immediate locations and the travelling in between. This may contrast with the experience of the separated parents who develop or continue to feel a more unitary sense of home (Haugen, 2007). It also involves changes in the daily or weekly rhythms of life; weekends and holidays may take on new meanings, being experienced in different environments with different sets of family members. These re-adjustments also apply to the parents and the new partners, providing an additional set of considerations to the existing organisation of timetables involving home and work.

The next is that these family practices involve widening and changing sets of relationships; not just the original parents and children but the new partners, possibly other sets of children, grandparents and other sets of kin. As has often been noted, these changes may involve subtractions as well as additions as when grandparents may feel in some degree excluded from the present arrangements. Put in terms of the key themes of this book, the 'others' who are taken to account in the performance of family practices will differ and this provides the novelty and complexity involved in these moves.

To some extent, the changes in time and space following separation and re-partnering may present themselves as constraints. There are new considerations to enter into the daily or weekly timetables and these considerations impinge upon both parents and children. Yet there may also be the opportunity to define and redefine relationships, and children may develop strategies for transforming imposed time schedules into their time in their places. For example, the child may develop new knowledge about the new neighbourhood and its inhabitants and may become an expert about the locality in contrast to the parent who simply visits on a regular or irregular basis. The apparent constraints of having to be in a certain place at a certain time may, over time, enter into everyday negotiations with significant others, adults and children alike. The apparent inevitability of timetables and scheduling will be shown to be more open to challenge than might first have seemed.

In one sense the temporal and spatial aspects of the family practices under these circumstances might seem to be both more diffuse and more weakly bounded. On a day-to-day basis, family practices may be carried on out of sight of former family members and without reference to them. But this, to some extent, will depend upon the nature of the social networks in the new household and its surroundings. More likely is the possibility that the family practices will often be more concentrated at certain periods of time (weekends, school holidays) when several sets of practices have to be tightly co-ordinated with each other and more diffuse during the times in between.

Generally, also, the re-partnering may lead to a more weakly bounded set of family relationships. At the very least, notions of 'the family as a whole' will become more problematic. More important, perhaps, is that the family boundaries will become weaker not simply in terms of notions of 'the family as a whole' but more in terms of the models of the family in the heads of different family members. Thus the family maps drawn by children may differ considerably from those drawn or imagined by other family members.

A second set of examples comes from a consideration of work-related mobilities and the way in which these impact upon the temporal and spatial practices of family members. Here a simple distinction may be made between low, medium and high mobilities.

Low mobilities

This is a situation where one or more partners are working at home. The temporal and spatial dimensions of family practices might become, initially at least, more explicit as negotiations are required around the divisions between work time and family (or non-work) time and space. These complications are enlarged when there are children present in the household and where adjustments have to be made according to the rhythms imposed by school. This entails a relatively high degree of concentration of family practices which have to be co-ordinated with each other in a way which is regularly visible.

It is likely that these issues of temporal and spatial co-ordination will increase in significance as more and more employees (perhaps for environmental reasons) are encouraged to work from home for at least one day a week. This may raise novel and unanticipated questions. Is a 'lunch break' different when taken at home from a lunch break at work which could be anything from a sandwich at a desk to a deliberate trip to a canteen or a café? Early retirement may also pose similar problems of spatial and temporal co-ordination when, for example, in more traditional households, men may express concern about not 'being under the wife's feet' (Cliff, 1993).

Medium mobilities

For a long time, this might be seen as the standard model of work/ family relationships. One or both partners tend to work away from home on a fairly regular basis, either on a full-time or a part-time basis. The complexities of combining and negotiating timetables, not only in relation to journeys to and from work but also in relation to opening hours and school timetables, have been well explored and constitute the core of what is sometimes referred to as 'work-life balance' (see Chapter 9 for further elaboration).

High mobilities

This is a situation, again like low mobilities possibly growing in significance, where one or more partners may be away for substantial periods of time. Traditionally, this has involved occupations like deep-sea fishing and the military (Chandler, 1991). In these cases we are dealing with a highly gendered division of labour where the husband is away for substantial periods of time and where there are clear distinctions between away time and home time. More recently attention has been focussed on the impact of employment in global organisations which require frequent travel abroad. Again these distinctions may be highly gendered, part of the construction of new globalised masculinities, although the tensions around notions of fathering might become more openly recognised. Brandth and Kvande have explored what happens when globalised, and largely masculine, careers meet government schemes for parental and paternal leave (Brandth & Kvande, 2001, 2002). Looking at these high mobilities in general we see a mixture of concentrated (during home periods) and diffuse practices (when one partner is away) and possibly a more weakly bounded set of family practices.

Clearly, the impact of these different mobilities will vary according to the life-course as well as with the actual set of persons within a particular household. (The discussion here has assumed two adults and some children but this obviously does not exhaust the possibilities.) Any given household may include a mixture of different mobilities such as where one partner works at home and the other goes to work on a daily basis or where one partner is away for long periods of time while the other works on a daily basis. We have reason to suppose that the high and low mobility patterns will increase in significance and may continue to have implications for the spatial and temporal dimensions of family practices.

A third set of examples to illustrate time/space and family practices can be provided by the impact of global migrations on family practices. The overall impact of global working practices has already been mentioned but here we are referring to patterns of migration and resettlement between and in different countries. Recent studies have reminded us of the complexities of the relationships between these movements and family practices and of the dangers of stereotyping the family practices of migrant groups (Erel, 2002). Clearly such movements have the effect of enlarging the space within which family practices are conducted and, as we shall see, also have an impact on temporalities as well. The members of transnational families are separated

> yet hold together and create something that can be seen as a feeling of collective welfare and unity, namely 'familyhood' even across national boundaries.
>
> (Bryceson & Vuorela, 2002: 3)

For a more extended illustration I shall take Jennifer Mason's analysis of 'the visit', whereby Pakistani families located in Britain arrange for regular returns to their place of origin (Mason, 2004). In this context, the term 'visit' has a specific meaning referring to a regular period of anything from two weeks to several months. These visits take place every, 2, 3 or 4 years and are the subject of planning and saving during the intervening years. (It should be noted that Mason's qualitative sample is weighted towards middle-class families with fairly high educational qualifications.)

Mason seeks to answer the question 'What is the visit about?' This might seem obvious but a detailed analysis of the responses reveals several layers of meaning.

In the first place, it is simply about visiting, a repeated activity that affirms and renews existing ties and is a form of display (Finch, 2007) stating that 'we are a family that works'. As such it has a lot in common with many other family gatherings where distance is not such an obvious factor.

Second, it is a process whereby family members may get to know their relatives: 'shared kinship biographies can emerge and be sustained over time, with an assured past, present and future' (Mason, 2004: p. 425). This is a pithy statement of the relationships between time and space in relation to the visit. This may be particularly important in the case of children who come to know relations who, up to the present, have been just names or photographs and, similarly of course, these relations come to know new children.

Third, and this is clearly related to family practices, the visit provides an opportunity for doing things together. These include visiting other sets of relatives, having meals, shopping and exploring neighbourhoods. The practices here are clearly *family* practices. Finally, the visit may provide opportunities for being there at key moments to do with the life-course or for clear reasons for shared celebration.

Mason does not ignore the risks associated with visiting which is more usually from the UK to Pakistan rather than in the reverse direction. These risks arise out of the fact that the visiting family members are in a permanent state of being visitors without a more regularised position within the wider family. There is to some extent a tension between values associated with hospitality and the values associated with family and kinship. Further, it is always possible to upset local sensitivities which may have been forgotten or unknown as a result of geographical distance. In his autobiographical work, Barack Obama recalls his visit to Kenya and how there are difficult decisions to be made about who is to be visited and in what order (Obama, 2007).

In these examples there is a sense in which time is constructed so that the visits become regular events which are planned for and anticipated. But it may also be the case that the felt necessity for making a visit may be seen as some kind of constraint. The visit is clearly an example of concentrated rather than diffuse family practices as large numbers of family members are involved and there is a considerable degree of orchestration of several different activities. At the same time it involves, if not a redrawing of family boundaries, a temporary shift in the range of immediate affect.

We might argue, in fact, that a consideration of the movements of transnational families in particular may enhance the value of a family practices approach since notions of 'the whole family' become more open to question (Erel, 2002) and fluidity and movement come more to the fore.

Conclusion

Spatial metaphors are often used to describe family relationships or to account for particular family practices. 'We are not very close' may, for example account for infrequent visits or invitations. 'Closeness' here is not determined by geography or genealogy as much by personal preference and the extent to which the two biographies are interwoven. Looking more specifically at the geographical dimension, most of the evidence suggests that there is a relatively weak association between geographical distance and emotional distance but, as the example of 'the visit' shows, the character of the relationship may be affected. Geographical distance may mean that visits have to be planned for in advance and hence there are new temporal considerations introduced.

This discussion has highlighted the reflexivity of time/space and family practices. It is not just that given family practices take place in locales defined by time and space (mealtimes, bedtimes, celebrations, visits). It is also that a sense of time and space is created or recreated by these practices and the relationships involved. Annual family gatherings provide reminders of the passing of time, reaffirmed by photographic memories. In the case of space, more specifically, family practices do not simply take place within homes but they also re-affirm the symbolic and imaginary sense of what home is about. When individuals state that 'a house is not a home', they are pointing to the close, reflexive interplay between space and family practices (Holdsworth & Morgan, 2005).

Another feature which is highlighted in this discussion is the significance of mobilities in everyday family life. These are a necessary corollary of the interplays between time/space and family practices. These may be simple everyday mobilities within the home: 'upstairs to bed' or 'outside to play'. They may be regular mobilities shaped by the relationships between home, work and community, journeys to and from work, the school or the shops. Or they may be longer-term mobilities between different 'homes' (as in the case of children in reconstituted families) or planned visits or family holidays. To provide an autobiographical example, my father used to take me for walks on a reasonably regular basis, sometimes in the countryside of the Home Counties and sometimes in London. Many of my memories of my father were of these movements through landscapes on foot or the train journeys that took us to the points of departure. Part of the character of this experience lay in the novelty of the landscapes and the items encountered through these movements. These movements contrasted with the shorter trips more usually conducted with my mother to the shops or to visit neighbours. The way in which family practices are also moving practices is relatively unappreciated.

The remaining question is whether the family practices approach provides for a greater appreciation of the relationships between time and space in family relationships. The contrast here is with statistical or demographic accounts or with more theoretical models. It is true that demographic tables necessarily include elements of time and space in them and this is made explicit in discussions of birth cohorts or household compositions. But it is also true that considerable work has to be done on the part of the reader to get a sense of time, space and movement from these tables and diagrams and that readers are differently skilled when it comes to appreciating these dimensions in what might otherwise seem to be static representations. Similarly many theoretical accounts – functional, exchange theory, rational choice – may have some spatial and temporal elements congealed within them but their initial appearance at least tends to be more static. The arrows that link different parts of a model, for example, may suggest a movement of sorts but it is at some distance from the experiences of the actors concerned.

There are two features of family practices which allow for a greater appreciation of time and space. The first is the emphasis upon doing. Deeds and activities take place in particular temporal and spatial settings. Throwing a ball can take place anywhere and at any time but only with the aid of a particular pair of spectacles can it become a family practice, something conducted in the park or the garden or on holiday between family members.

The second is the stress on fluidity. The very notion of fluidity conveys a sense of movement and spatiality even hough, in this case, the term is used metaphorically. The fluidity is a recognition that what may be defined as family practices might also, with another set of lenses, be seen as gendered practices, working practices, health practices or whatever. Nevertheless this provides for a recognition of the spatial and temporal locations of family practices but one which recognises that these practices are part of a much wider and larger set of complex social practices.

It can be concluded that, potentially at least, there is a much greater affinity between the practices approach and a more nuanced recognition of the necessarily temporal and spatial character of these practices. Indeed, as with a discussion of home (with its connotations of public and private) or with a discussion of the life-course, an understanding of family practices may take us closer to understanding time and space themselves. At the same time, reciprocally, a recognition of the centrality of temporal and spatial considerations may take us closer to an understanding as to what family practices are about and an appreciation of their essentially everyday quality.

6
The Body and Family Practices

Introduction

The body, in common with time and space, has become an increasing focus of attention in sociological enquiry. Indeed, these three concerns are closely linked since bodies necessarily must be seen in time and space. However, some of the literature on the body appears to have little connection with families or, indeed, other collectivities or relationalities. As I noted in 1996, family sociology and the sociology of the body appear to have few points of intersection (Gabb, 2008; Morgan, 1996). Where issues of children and childhood are discussed in terms of embodiment (e.g. Prout, 2000) the relevant sites are more often than not outside the home, in the school or playground for example. There is, perhaps, a danger that the body may be understood in individualistic terms through the connections between discussions of the body and the phenomenology of the self.

However, starting from the perspective of family practices it can be seen that the body is necessarily a relational body and this sense of relationality can be best understood through a distinction between the notion of the body and the process of embodiment. Whereas to talk of the body is to talk of separateness and distinction – my body and your body – the idea of embodiment is necessarily relational. Embodiment refers to an individual's sense of having or being a body and the apprehension of the other as being similarly embodied. Modesty, for example, is a response to the gaze, or the imagined gaze, of the other. This relationality may be extended to the material world in general and not simply to other social actors as, for example, when I trip on a loose paving-stone or cross to the more illuminated side of the street.

This sense of embodiment, of having or being a body, is not a constant sense. Rather we can imagine a continuum ranging from family practices which are relatively disembodied to those which are highly embodied. Those at the latter end of the continuum are probably easier to grasp. A strong sense of embodiment may be associated with sexual activity, with giving and receiving care and with violence between intimates. Relative disembodiment on the other hand may be seen where account is taken of family others even where these others are not physically present. Urry's discussion of the different dimensions of co-presence (referred to in the previous chapter) is relevant here; co-presence can also be imagined or virtual (Urry, 2002). Other family members may be influential and taken into account across time and space, across generations and, indeed, across the limitations of an individual life-span. One does not necessarily have to sign up to the details of a Freudian analysis to recognise the influence of significant others who have recently, or not so recently, passed away.

I have used the phrase relatively disembodied here since it is likely that some sense of the other as an embodied family member will persist either through memory, through abiding visual representations in the form of photographs or through real or imagined family resemblances in other family members who are actually co-present. Immediately when someone takes account of another family member, even where that other is not physically present, that other is understood in embodied terms, even if that sense of the other does not necessarily correspond to how that other is actually looking at the present time.

This sense of embodiment, family embodiment, is clearly linked to the life-course. This is partly because individual bodies change over time through the processes of growth, maturation and decline, even although the links between physiological changes and their personal understanding and meaning may not always correspond. Thus the meanings of 'putting on weight' or 'becoming bald' will be negotiated within family practices. But it is also because within the life-course there are different periods of bodily intensity associated with, for example, infancy or old age, puberty or the early years of marriage. Again it should be stressed that these periods of intensity, focussed around different forms of physical care for example, are not straightforwardly determined by 'the stages of the life-course' but are framed within the meanings associated with particular family practices.

Another way of thinking about changing patterns of embodiment over time, associated with the life-course, is in terms of different configurations around the family 'we'. Any sense of relationality is a sense of a 'we' although the configurations that make up this 'we' will shift over time. It is important to recognise that this sense of a family 'we' is both relational and embodied. The family members who constitute the 'we' at any one point of time are embodied others who are spatially located.

What is becoming apparent (although it should have been apparent a long time ago) is questions of embodiment do not constitute some optional extra in family analysis, a modish addition to spice up what might otherwise seem to be routine accounts, but constitute the core of what family practices are actually about. Indeed, the very notion of practices may be defined in bodily terms as Reckwitz notes when he writes that 'practices are routinized bodily activities'(Reckwitz, 2002: p. 251). He goes on to elaborate that 'a social practice is the product of training the body in a certain way' (ibid.). Clearly, for example, Goffman's work is full of examples of the relational body, of the 'positioning of the body in social encounters' (Giddens, 1984: p. xxiv) although this discussion is not focussed on family relationships.

It should come as no surprise, therefore, that issues of the body and embodiment are central to the analysis of family practices. Jacqui Gabb's recent book on researching intimacy in family (Gabb, 2008) provides numerous examples, some of which can be illustrated by this open-ended list:

- Nudity between family members, and negotiating boundaries between the public and the private.
- The design of domestic space and how this shapes or reflects embodied family practices.
- Expressing emotions between family members.
- Beliefs about and practices of corporal punishment.
- Routine and non-routine practices of care between family members.
- The position of domestic animals within routine family life.

The list could be extended. In the rest of this chapter, I shall look at some of these, and other, family practices in a way to explore issues of embodiment further. I begin by looking at some more general considerations to do with the family gaze, embodied knowledge and bodily density before using these to examine more specific considerations. These considerations are to do with care, food and feeding, violence and embodiment beyond the life-course. In the concluding section I shall ask whether consideration of embodied family practices raises any special issues or whether the family is simply one site among many where embodied practices take place.

The family gaze

We recognise, evaluate and interact with others through our senses, frequently if not invariably or inevitably through the sense of sight. For those sighted members of the population, the majority, it is through seeing the other that we make distinctions between strangers, acquaintances and intimates and respond accordingly. But seeing is not the same as looking and this latter term is surrounded with a whole host of political or ethical considerations. Children are told not to stare and some male gazes may be read, disapprovingly, as ogling or leering.

The term 'gaze' entered the social sciences and cultural studies some years ago ('the tourist gaze', 'the male gaze', etc.) and when the term is used it is clearly referring to looking rather than seeing. It is referring to a kind of relation of the observer to the other in such a way that the former consumes or constructs the other within a particular frame of reference. Implied here is a lack of reciprocity between the gaze and the gazed-upon.

The term 'the family gaze' has been used in the literature (Haldrup & Larsen, 2003) although the usage here is an extension of the idea of 'the tourist gaze' and refers to the ways in which the latter is something frequently constituted by families on holiday rather than by isolated individuals. My usage will depart from this and will focus on the way in which family members constitute other family members or relationships through the deployment of the gaze.

The significance of the family gaze in the sense that I am using the term lies in the frequent co-presence within relatively bounded spaces of family members of different, ages, generations and genders. It is not simply that the external bodily features of the other (height, weight, colour of hair, posture and so on) are regularly and routinely available to others but that these immediate bodily appearances are frequently taken as signs of 'inner' states, anxiety, displeasure and so on. Further, to be able to conduct these readings from bodily appearances may be seen as manifestations of caring about the other or, at least, an acknowledgement that a relationship exists. The fact that misreadings may occur does not detract from the fact that they take place and on a regular basis.

There are some further points to be made about the family gaze. The first is that there is frequently some kind of modification of the general prescription (allowing for cultural variations) against staring, that is, against too prolonged gazing upon the other. There may be some lack of reciprocity here such that parents may gaze with more frequency, and will claim the parental right to do so, on their children than the reverse.

However, the bodily monitoring (in questions to do with health, for example) is never purely one-way. Backett-Milburn's discussion of health practices within middle-class families shows children monitoring the health and bodily appearances of parents and vice versa (Backett-Milburn, 2000). Between intimate partners, looks may constitute the equivalent of a turn taken in a conversational exchange or may be the object of some negotiation. 'Why are you looking at me in that tone of voice?' goes a popular joke and a silent, particular kind of look may be responded to with a 'What?' Again, intimates may claim greater licence in taking pleasure in the other's bodily appearance than would be the case in relation to strangers or acquaintances.

One interesting extension of the idea of the family gaze is to be found in Hochschild's discussion of children as 'eavesdroppers', an idea that deserves more development (Hochschild, 2003: pp. 172–81). Children frequently have an ambiguous status within the household being sometimes fully constituted actors while, at other times, unrecognised observers or even, temporarily, non-persons. The occasions where this eavesdropper status may be heightened are those where adult concerns and adult projects (financial concerns, separation and re-partnering, negotiations about divisions of labour) come to the fore. Here, the adults become the object of the child's gaze and the young person may become a kind of lay sociologist or psychologist observing and interpreting adult behaviour. A good fictional illustration is provided by Henry James in his *What Maisie Knew* where Maisie is the child who observes the complex and shifting world of adult relationships.

Thus the family gaze represents a particular form of family practice brought about by the physical co-presence of family others within a relatively confined setting. This can include others who make regular visits to 'the family home' as well as those who, for varying periods of time, occupy that home. It should be stressed that while the term 'gaze' refers primarily to the sense of sight, similar principles apply where this is not possible or in relation to other senses such as smell, hearing and touch. Indeed, touching, smelling and tasting the other constitute part of the privileged access that a sense of intimacy implies. It is not simply that family members can 'see' other family members but that they can look at these others in particular licensed ways. The deployment of these gazes will sometimes be the subject or sanction or negotiation but the fact that these negotiations can and do take place is part of the idea of the family gaze as a particular form of practice.

One everyday example of the working of the family gaze and its relationship to embodiment is to do with matters of physical growth.

James notes how phrases, in relation to children, such as 'hasn't she grown?' or 'when you grow up' mark the embodied passing of time or future developments (James, 2000). While her focus is on children's bodies outside the home (in the playground, for example) it can be readily adapted to family practices. To provide an autobiographical example, my father regularly marked my physical growth with pencil marks on a door frame and, later, note was made of the time when I threatened to become taller than him. The display of children to infrequent visitors to the home often involves statements about growth, focussing initially on the more obvious embodied indicators.

One particular extension of the idea of the family gaze is through the regular use of photographic images, ranging from carefully framed and publicly displayed representations to images (still or moving) on the Internet or on mobile phones (Davies, 2010). Much can and has been written about these representations. Here I want to highlight the taking and display of photographs as a particular form of family practice and family display (Finch, 2007). I want to point to the embodied nature of this activity and the way in which it constitutes family relationships. Imagine, for example, two or three related persons going through a photo album. Confronting a particular representation, someone may say, 'that is just like her'. This is not, I would suggest, simply a physical recognition of the person depicted, our mother say, but also some kind of claim about the relation between the physical appearance, (the smile, the hair, the stance) and the person's 'inner' self. It is a relational claim, reflecting on the relationships between all the persons concerned and the women depicted in the photograph. The family gaze, therefore, does not require the actual physical co-presence of other family members and may, indeed, cross the life span itself.

Embodied knowledge

I have argued (pp. 35–6) that intimacy consists of several dimensions and that one of these dimensions is a form of knowledge of the other which is distinct from the kind of knowledge that we may have of strangers or acquaintances. This is the knowledge that arises out of the interweaving of biographies over time and while family relations do not provide the only site where this takes place, family practices involve the use, the recognition and the generation of this knowledge. Much of this knowledge may be described as embodied knowledge.

To a large extent, the knowledge that we have of any other person, whether that person be an intimate, an acquaintance or a stranger, is embodied, especially if when we use the word 'embodied' we also include things like tone of voice. This embodied knowledge can, of course, extend to celebrities or fictional characters. What, then, is distinctive about the embodied knowledge that we have of intimates, especially family intimates? The following list, although incomplete without doubt, should be enough to indicate that there are some important differences:

- The embodied knowledge is simply more detailed and covers a wider range of possibilities in terms of things like facial expressions, bodily demeanour and gestures. Further, as I suggested in the previous section, these embodied manifestations may be routinely monitored in order to gain insight into emotions and how the particular intimate relation stands at the present time.
- The embodied knowledge has a detailed history, the consequence of sharing space, time and experiences. This means, for example, that we can remember previous bodily characteristics such as before I went bald, when I had a beard and so forth. In some cases, of course, this may go back to the time of birth and early infancy.
- The embodied knowledge is not limited to the senses of sight and hearing. Touch and smell may be especially important. Goffman, for example, shows how people frequently present an intimate relationship through the deployment of touch (Goffman, 1971), holding hands for example.
- The embodied knowledge ventures into certain tabooed areas. With very few exceptions, the legitimate knowledge of the other's naked body is confined to intimate relations. The knowledge of the other that is obtained through the sense of touch also refers to practices that venture into these tabooed areas. However, with taboo comes risk and, especially where the relationships between adults and children are concerned, the perceived element of risk frequently comes to the fore (Gabb, 2008).
- The embodied knowledge frequently is associated with particular locations within the home such as the bedroom, the bathroom or the toilet.

Further aspects could be developed but I hope that enough has been said to indicate that the embodied knowledge between intimates is of a different order from the embodied knowledge that we routinely gather and use in everyday life. This knowledge may not be confined to family practices, conventionally understood. For one thing there are cultural variations and for another we may see these as applying to sexual partners outside family relationships or, in some cases, to friends. Nevertheless this embodied knowledge within family relationships and practices has an over-determined or intensified character. It is not that the examples of embodied knowledge that I have provided, examples which come from a wider set of possibilities, can be found individually within family practices but that many of them together may be so discovered. While, for the purpose of analysis we may list embodied knowledges, in practice they often merge together to constitute a coherent sense of a particular set of family relations.

Bodily density

Our sense of embodiment is not, I have argued, a constant or regular feature of everyday life. There is considerable variation according to the extent to which we are brought into the proximity of other embodied selves and the meanings attached to these different clusterings. There are some temporary clusterings (standing-room only on the bus, an overcrowded lecture hall) which may produce some momentary discomfort and annoyance but little more than that. There are other temporary densities which may be experienced as enjoyable and as contributing to the core meaning of the event in question such as discos or sports events, either as players or spectators. Similarly a lack of bodily density may be experienced as pleasurable (a walk in the countryside) or as threatening (empty streets at night).

When we think of bodily density in relation to families, it is likely that notions of overcrowding may come to the fore. It will be remembered that the concern with overcrowding was not simply with a number of people in relation to the space available but with the extent to which males and females of different ages were sharing the same space and the extent to which different functions (eating, sleeping, washing, etc.) were similarly condensed. There is clearly considerable historical and cultural variation as to what is seen as acceptable in these matters.

From the point of view of thinking about bodily density, the embodied aspect of family practices, we miss the point if we concentrate solely on over-crowding. While this may still be an important consideration in many parts of the world, the concerns here are somewhat different. The differences emerge when the dimension of time is included. The bodily density of family practices emerges out of the fact that a relatively small number of people see and have to take account of each other for relatively long periods of time. This length of time may be measured in terms of the life course, for example in terms of the number of years it takes to bring up and 'launch' a child into the outside world, or in terms of the number of hours per day. Of course, individuals move in and out of this relatively dense circle and, over time, the 'family home' may become a place that is visited from time to time rather than where one resides on a permanent basis. These domestic clusterings may be augmented by friends and neighbours.

This domestic bodily density does not require the constant co-presence of all the others all the time. Movement in and out is, as we have seen, built into everyday family living. Nevertheless there are constant reminders of these others in their absence such as particular chairs, clothes hanging in a wardrobe or a book on a table. Gabb refers to 'the spatial messiness of everyday domestic living' (Gabb, 2008: p. 88) and it is this messiness that provides a sense of singularity and shared ownership of domestic space. My sense of bodily density does not require the constant presence of all who might be defined as 'belonging'; all that is required is that some people are present in different combinations and for different periods of time and that there are these embodied traces which provide reminders of others who have been there and who will return.

This notion of bodily density reinforces the processes discussed in the previous two sections, the idea of the family gaze and the embodied knowledge that family members have of each other. The possibilities for mutual monitoring, perhaps surveillance, and for the accumulation of embodied knowledge are greatly enhanced through this fact of bodily density. Much of the knowledge that intimates have of each other is not obtained through conversational exchanges but through simply sharing space.

There are two further points that need to be made about bodily density and family practices. The first is that the density is not a constant but something that varies over a life course or as a consequence of processes such as divorce and separation or death.

The readjustments that have to be made once children 'finally' leave home has often been commented upon. At what point, for example, does a daughter or son's bedroom become a spare room? Retirement, on the other hand, may bring about an increased sense of bodily density, well expressed in phrases about being under each other's feet.

The second is that bodily density is not necessarily experienced as something which is oppressive or unpleasant. It can be, of course, especially where abuse, violence or the threat of violence is present. This sense of density is reinforced where the person at the receiving end apparently has nowhere else to go. But bodily density may also be a source of the most intense memories and part of the process of establishing and maintaining ties even where, as is often the case, there is some measure of ambivalence. These might be memories of family members crowding around a meal table or sleeping on the dining room floor on festive occasions. These experiences become part of the stories that are told to others and these stories themselves are important family practices.

Bodily care

Discussions of care, for the most part informal and unpaid, are closely associated with discussions of family practices. Feminist scholarship greatly enhanced our understanding of the range of meanings attached to this theme and provided an appreciation of how family practices in this respect at least were also gendered practices. An important development was the realisation that care is a relationship and the practices and meanings on the part of the cared-for are as significant as those attached to the carer (Morgan, 1996: pp. 95–112). A further important development, especially relevant here, is the realisation that caring practices are embodied practices.

This is true whether we are talking about 'caring for' or 'caring about'. While the practices associated with the former are more obviously embodied, 'caring about' is frequently expressed through the deployment of the body whether this be through touching or embracing, through particular forms of eye contact or through some alteration of the tone or level of voice.

However, it is caring for which is the most obviously embodied and this is a point that does not need to be laboured. We also have some detailed understanding of the processes by which this form of care comes to be especially identified with the family and the more immediate processes by which a particular individual, often a woman, becomes the care-giver within a particular family circle (Finch, 1989; Finch & Mason, 1993). Rather than enumerate the ways in which this family-based care is embodied, I shall consider one particular case which explores the complex ambivalences that arise when you get a 'combination of emotion work and intimate hands-on care' (Isaksen, 2005).

We have a sense of the embodied nature of this discussion from some of the words in the title of Isaksen's article: 'dirt and disgust'. In this Norwegian study, she is dealing with matters that are often marginalised in discussions of caring. Her case deals with Brenda who is a widow in her mid 40s. She works full-time and lives with her teenage daughter but also cares for her elderly mother and her father who live nearby. Her father 'suffers from severe physical balance problems and incontinence'. A couple of brief quotations from Brenda will show how these issues of care provide a lens through which numerous issues of family care and embodiment are focussed and given an intensity through their combination:

> I feel my brother has betrayed me ... leaving all the responsibilities to me ... Even if he can't give any practical help, be could at least give me his emotional support.
>
> I have had a hard time to make ends meet. And then I have my daughter to take care of and all the troubles with my father and all on top of it ... Some weeks ago I nearly crashed my car two times within a few days ... Then I understood that I was too stressed and tired of it all.
>
> But now ... it's difficult for him[her father] to deal with the fact that he can't go for a walk without help ... He just can't accept it ... and he takes it out on us ... My mother has to take the brunt of it ... But still ... it doesn't make it easier for any of us.
>
> (Isaksen, 2005: p. 117)

These extracts from some longer quotations highlight the embodied nature of care and the way in which these reflect and are reflected in ongoing sets of family practices of some complexity. There is the stress on the fact that the care for the father requires a lot of highly physical hands-on work. There is the impact of these caring practices on this particular carer and on the sense of exhaustion and tiredness that this brings (see also Widerberg, 2005). There is the intermingling of emotional and physical labour and the multiple work and family obligations that Brenda has to juggle. And there is the ambivalence that lies at the heart of many family-based caring practices; elsewhere Brenda states that 'he has always been an emotionally difficult person' (Isaksen, 2005: p. 117).

In terms of the ideas developed earlier in this chapter we can see how the idea of the family gaze can be applied. Here we are reminded that the gaze may sometimes be required or expected of a family member who is in the position of a carer. Brenda feels obliged to monitor the conditions of her aging parents although, one assumes, this gaze is not always to be welcomed as it serves as a silent reminder of the cared-for's incapacities. We can see the embodied knowledge that Brenda has of her parents and of others to whom she is connected and this knowledge is in part the outcome of the bodily density within her family setting. Here this sense of density is not necessarily dissipated as a consequence of her living separately from her parents. Bodily density, as I pointed out, is a matter of time as well as of space.

It is possible that I am reading more into this short case than is perhaps justified. However, even these few extracts highlight the complexities involved in the caring relationship and the tensions and ambivalences arising out of the mixture of physical and emotional labour. Most important, we can see the embodied character of caring within families in terms of what is required on the part of the carer (and the cared-for) and the embodied experiences of the carer herself.

Food, feeding and family practices

Closely connected with issues of care, and equally gendered, are issues to do with the most fundamental of human activities, the preparation and consumption of food. While not all feeding practices take place within family relationships, there is undoubtedly a close affinity between food and families. The buying and preparation of food is frequently carried out, not simply in terms of one's own tastes or preferences, but also with references to the tastes, preferences and diets of others and these others are frequently other family members. Care (caring for) is demonstrated in the way in which food is prepared in terms of these other reference points or in the way in which implicit reference is made to the health and well-being of partners or children. Food is central to almost any family celebration or form of family display. And the notion of the family meal, and its alleged decline, remains prominent in representations of family life (Jackson, 2009).

What all this means is that it is not simply the case that feeding takes place, to a large extent, within sets of family relationships. It is also true to say that feeding provides one of the major ways through which family is constructed and reconstructed (DeVault, 1991). DeVault makes the useful distinction between 'cooking' and 'feeding', the former referring to the techniques involved, the latter to the many ways in which the expectations and needs of others are taken into account when planning meals. Her study highlights the complexities involved in this most everyday, yet fundamental, of activities and while feeding clearly can involve non-family members it remains the case that there will always be close relationships between family practices and feeding practices. And, where other non-family members are involved in a particular meal we may find some form of family display taking place.

And yet, increasingly, it is becoming apparent that our frame of reference cannot be confined to those described as family members. Food and feeding, it would seem, are too serious to be left to families. While much the same can be said of other sets of family practices, the implication of food practices within the welfare regime triangle of family, state and market provides, perhaps, the clearest illustration (Hobson & Morgan, 2002: pp. 9–10). Our focus is clearly on the intersections between food practices and family practices but it is clear that the market cannot be excluded from this discussion. The market is the source of most of the food that is prepared and consumed within the family and, consequently, the source of much that is seen as healthy or, more frequently, unhealthy in our everyday food. Moreover, the market is the source of many of the representations of the links between families and food. Similarly, the state is increasingly implicated in a whole range of feeding practices that impinge upon or involve family members. Much of the focus is upon healthy eating and children although campaigns about eating fruit and vegetables ('five a day') and exercise are directed to all family members (Jackson, 2009). The state is also concerned with the practices of food providers.

A clear example of how this welfare regime triangle impinges upon food practices is to be found in current concerns with obesity. Debates often begin with some statement of the scale of the problem:

> Almost two-thirds of adults and a third of children are either overweight or obese, and work by the Government's Foresight programme suggests that, without clear action, these figures will rise to almost nine in ten adults and two-thirds of children by 2050.
>
> (Ford & Fraser, 2009: p. 20)

While some attention is directed to food providers and to the high salt or sugar content of some manufactured foods, a great deal of the focus is on food practices within families. As the mother is perceived as being the person with the greatest responsibility for the health and welfare of children, several government interventions focus on her. In some cases there is recognition of the fact that social inequalities and class are important mediating factors.

The concern, therefore, is with the way in which family practices construct, or are seen as constructing, bodies, in this case the obese body. It is hardly possible to ignore the strong moral dimension to these debates and campaigns and the ways in which they may be seen to be concerned with the allocation of blame. This moral dimension is hardly absent from wider discourses about food and feeding. Healthy foods are 'good' foods and correct feeding patterns are identified with responsible parenting. The close linkages that are established between care, morality and family practices (frequently filtered through a gender lens) are clearly demonstrated in current concerns about healthy eating.

The associations between feeding practices and family practices also highlights the way in which, as argued throughout this book, family practices overlap with other practices. DeVault's study (1991) clearly highlights the strong overlaps between family practices and gender practices. This is reflected in the subtitle of her book: 'the social organization of caring as gendered work'. We see this not only in the routines of provisioning and feeding but also in the passing on of culinary skills across generations from mothers to daughters. This study also shows the overlaps between family practices and class practices in the business of feeding the family, a theme which is emphasised in numerous other studies (Jackson, 2009).

Violence

Issues of violence and abuse within the family have gained increasing recognition within the last thirty years or so, partly as a consequence of feminist scholarship. We are now much more aware of both the extent of violence and abuse within families and of the range of practices that can be considered. Initially, the concern was with violence against and abuse of small children and this has extended to consideration of violence against women partners and on to violence against men partners and the elderly within the home. Yet again, the issue of rape within marriage has been put on the agenda and has been the object of legal intervention.

The first point to emphasise here is that we are clearly dealing with matters to do with bodies and embodiment. At one sense this would seem to be obvious. There can be few acts more obviously embodied than striking or sexually abusing another person. Similarly, neglect of another, a child or an elderly person, has direct physical consequences. But matters go further than this. Emotions to do with anger or physical desire have consequences for the perpetrator as well as for the victim. More than this we are dealing with violent or abusive relationships so that both victims and perpetrators come to see themselves and each other in terms of a shared, if not mutual, embodiment.

Some of the themes introduced earlier in this chapter may illustrate and develop this point further:

The family gaze

Persistent patterns of violence or abuse shape the family gaze in particular ways such that the other is looked upon as a potential source or object of further violence or abuse. Again, the question of time becomes central. It is not the present act of violence which is alone of bodily significance; past acts and the expectation of future acts are also part of this. Autobiographical accounts, for example, frequently include a child's sense of apprehension as the father returns from a night at the pub. This is what distinguishes family violence from other more public forms of violence, such as muggings. Further, within the family setting there may well be spectators to the violence between a parent and a particular child or between partners.

Embodied knowledge

Experience of violence or abuse is a particular form of embodied knowledge and one which highlights the boundaries between the public and the private. The secrecy surrounding the violent or abusive episode or the patterns of denial or disavowal following that episode becomes part of the violence itself. The phrase, 'behind closed doors' readily conveys this sense of abuse, secrecy and public denial (Gabb, 2008).

Bodily density

Experiences of abuse and violence serve as direct reminders that issues of bodily density are not to be confined to discussions of overcrowding but more to the way in which the constant physical presence of others serve as reminders of relationships which are experienced as oppressive and exploitative. Important here, also, is the perceived lack of escape on the part of the abused child or partner and, indeed, it has been argued that the lack of ways of escape may even affect the perpetrator of a violent act (Goode, 1971).

Thus the embodied nature of domestic violence or abuse goes beyond the sheer physicality of any individual act but extends into memories of the past and anticipations of the future. The experience of such acts provides the strongest illustration of the understanding of embodiedness that is being presented here. This is not simply a sense of one's own body and its boundaries but a sense that this is essentially related and bound up with the practices of others with whom one shares domestic or intimate space.

A remaining set of questions are to do with the strength and the directness of the connections between family practices and domestic violence. To what extent are these distinct and, possibly, built into family practices? The discussion in terms of the family gaze, embodied knowledge and bodily density may provide some reminders of the way in which particular family constellations may contribute to the character of these violent relationships even if they may not have a direct causal effect. To explore this further I consider two different dimensions which shape the ways in which violence is understood.

The first concerns the extent to which the violence is legitimated or otherwise. To be absolutely clear, note that I am talking about 'legitimated' (which refers to a social process) rather than 'legitimate' (which refers to a stable moral status). All kinds of acts which might formally be described as 'violent' in that they involve some unwanted physical invasion of an other are, in fact, subjected to processes of legitimation which, in effect, denies the label 'violent' itself (warfare, sport etc). Most relevant for the present discussion is the physical punishment of children within the home. Most research within the UK suggests that physical corporal punishment within the home is almost universal (Gabb, 2008: p. 83) although the extent and nature of this punishment varies considerably. To a large extent these are seen as legitimate family practices and any attempt to control this through legislation is resisted strongly. In the case of abuse, on the other hand, matters are much less straightforward where there is considerable uncertainty about where the boundaries between abusive and non-abusive parenting practices lie (Gabb, 2008: pp. 82–3). To give a simple example, should a parent take a photograph of a small child in a bath?

The second concerns the extent to which the violence is seen as understandable or, alternatively, as being beyond any reasonable frameworks of understanding. Some acts of violence may be seen as understandable, if not necessarily approved: a momentary loss of temper, for example. Other forms of violence or abuse might be seen as completely outside understanding as being, in effect, 'beyond belief'. This applies to a great deal of child abuse within the home. In Britain, the fairly regular publicity that surrounds particular events of violence, extreme neglect or abuse of children within the home (and the intervention or lack of intervention on the part of the police or social services) serve as occasions for the evaluation of the boundaries between the public and the private and the nature and limits of parental rights and responsibilities.

Putting to one side these particularly dramatic and distressing cases, it can be seen that family practices may be involved in making complex moral judgements about 'less serious' forms of violence. Thus family members may take part in discussions about provocation or about whether the violence is really abusive or whether it can be excused given some weighing up of plusses and minuses in relation to the perpetrator. Here the bodily density, the close interweaving of biographies over time, provide for the possibility of complex moral evaluations within the site of the home. Considerations of violence, abuse and family practices also raise questions of risk in the context of family living. This is not simply to reiterate the point that the risks of being on the receiving end of violence or abuse are strongly associated with family living rather than the more publicised locations outside the home or the threats from strangers. It is also, as we have seen, that parents (in particular) may be increasingly wary about crossing the line, or being perceived as crossing, the line between appropriate and non-appropriate forms of behaviour.

Embodiment beyond the life course

There is one final aspect of our exploration of family practices and embodiment, one that is not fully encompassed by the themes developed so far. This is to do with embodiment beyond the life course or, indeed, any individual life span. We may be concerned here with family practices prior to conception or, at the other end, family practices beyond death. These are fascinating questions and it is only possible to provide some brief indications of the themes involved.

One of the most famous opening lines in English literature is the beginning of Sterne's *Tristram Shandy:*

> I wish either my father or my mother, or indeed both of them, as they were in duty both equally bound to it, had minded what they were about when they begot me.

While an individual life-course is conventionally described as beginning at birth, the aspirations and practices of the parents take this trajectory back to conception and before. This is not simply a question of the inheritance, real or imagined, of particular physical or mental characteristics. We are also thinking about, for example, of the reproductive intentions of the parents and whether a particular pregnancy occurs after several unsuccessful attempts to have a child. Alternatively we may consider the child who learns, perhaps in adulthood, that he or she was an 'accident'.

The complexities associated with sperm donation provide insights into this extension of the life-course and, in addition, the extension of family practices to include others who are unknown or imagined (Tober, 2001). The extent and the ways in which would-be donors are profiled according to ethnicity, social class, education and other characteristics allows the social parents to elaborate imagined social relationships and the continuities of certain characteristics. The interplay across an individual life course may be seen where, in later life, there is some form of identity release and it becomes possible for the donor to become known.

The continuity beyond an individual life-course is also shown in considerations as to what happens to the body after death and the way in which this reflects family practices and beliefs. Brandes (2001) provides an intriguing case of a Guatemalan who died in the US and was, through an administrative mix-up, cremated. Such an occurrence created considerable distress among his community and wider family network at home where stress was laid on the importance of burying an intact body. This concern was expressed despite the recognition that the deceased led a life which was far from exemplary and further complications arose as claims for financial compensation were made and met.

Why is there such concern about bodies ending up in the wrong graves? Or about the bodies of soldiers missing in action? Or why might there be attempts to distinguish between bodies buried in mass graves? We may begin to respond to these questions by stressing that family practices do not end with the death of an individual. Relationships between the deceased and family others continue and these examples show that this is not simply a question of memories but of embodied notions of a proper death.

Conclusion

What does this discussion about family practices add to the increasing wider sociological interest in the body and embodiment? And what do considerations of the body contribute to the discussion of family practices? I would argue that the fact that discussions of the body and embodiment have arrived relatively recently within social sciences and the humanities is not to encourage dismissal of these developments as something fashionably and possibly, therefore, transient. The same might be said of discussions of time and space in the previous chapter. But, as with time and space, the question is not so much why we should begin to take account of these dimensions within social enquiry but, rather, why they were relatively marginalised for so long. And, again as with time and space, I shall argue that the connections between the family and embodiment are especially intimate.

At one level, family practices might be seen as being located at the core of any discussion of bodies and embodiment. This is because so much of what people talk about in connection with everyday family living so clearly relates to the body: birth, growth and maturation, sickness, disability and their management, care, sexuality and death. Earlier statements about the relationships between 'the family' and these processes probably contributed to a wariness about presenting too essentialist an understanding of family living; the idea that the family is a basic institution because it deals with these basic processes. But more sociological understandings have underlined the fact that the associations between these processes and everyday family can only be explained by detailed and complex historical and sociological investigation.

The emphasis on 'embodiment' rather than 'the body' highlights the fact that we are dealing with complex interactional, culturally and historically situated, processes. Discussions of the workings of the family gaze, embodied knowledge and bodily density can only hint at some of these complexities. What a focus on family practices provides is a clearer understanding of the relational character of the body and embodiment. As we have seen in the previous chapter, this relationality also, inevitably, brings in questions of time and space so that family practices may be seen as particular constellations of embodied activities temporally and spatially located. This means that the relationships are not simply with other human beings but also with domestic animals. Gabb (2008) devotes several paragraphs to discussions of family pets and the physical organisation of the home. Reference to embodied processes to do with feeding, sleeping and washing also refer to distinct locations within the home, which have their own constructed associations with notions of privacy, the public and the private, and divisions of labour.

So an understanding of some of the complexities of family practices can make important contributions to a relational understanding of the body and embodiment. In looking at the reverse relationship, the impact of studies of embodiment on the analysis of family practices, we can similarly argue that there is a genuine enhancement of our understandings. Family practices are practices which are carried out with reference to others who are defined as family members so that family ties are constructed and reconstructed through the enactment of these practices. The apparently innocent phrase 'with reference to' can include some directly embodied activities such as feeding, physical care or physical punishment. But it can simply mean remembering, taking account of, anticipating. While these are in some ways mental processes, operations that are often invisible to the other's gaze have embodied features. Taking account of may mean constructing a mental image of the embodied other family member and may have embodied consequences (anxiety, desire, pleasure) on the part of the person taking account. In all kinds of ways, therefore, considerations of the body and embodiment are intimately connected with family practices. While this is not exclusively true for family relationships (I may conjure up similar images or have similar bodily sensations when I take account of colleagues at work or neighbours) the exploration of processes such as the family gaze, embodied knowledge and bodily density certainly contribute to a particularly strong way in which embodiment is located in family practices and family constellations, practices and constellations that can exist outside an individual life course.

7
Emotions and Family Practices

Introduction

The study of emotions, as with some of the other topics discussed in the previous two chapters, arrived rather late within sociological enquiry. In part this may reflect an overall concern with rationality or a suspicion that to take emotions seriously would be to sign up to some form of psychological or biological essentialism. Even where emotions did enter into sociological study it would be, so some critics maintained, in such a way as to hold emotions at a distance. Emotions were seen as socially constructed or the emphasis was on 'emotional labour' rather than on emotions as such (Craib, 1995). It may be, however, that use of terms such as 'emotional labour' simply recognises that sociology does have a contribution to make to the study of emotions but that it is not attempting to colonise this particular piece of territory. Such usages may in effect suggest a recognition of the limits of sociological enquiry.

Putting these debates to one side for the moment, it can be readily recognised that emotions are a key element within everyday family life. Take, for example, this quotation from an American woman talking about feeding the family:

> I'm sure – I remember quite clearly as a child, and even to a certain extent now – texture of food is very important as to how you like it. And I would assume she doesn't like the texture of rice, because she likes noodles, and they slide down.
>
> (DeVault, 1991: pp. 87–8)

At one level, the talk is about some very practical issues to do with feeding children and the assumptions about children's likes and dislikes.

Yet words such as 'like' and 'dislike', although here directed towards different kinds of food, are to do with embodied emotions. Memories of having to eat certain disliked foods may remain for a long time and it is frequently difficult to disentangle the remembered feelings about certain foods from the family relationships within which these experiences were embedded. Similarly, the mother in this quotation is drawing upon her own childhood memories as a guide to her own feeding practices. Yet again, the routine planning of feeding and daily menus involves the anticipated reactions of family others and expressions of pleasure or otherwise.

Our very understanding of family practices, practices which are orientated to other family members and through which a sense of family is created and recreated, would seem to have emotions at the heart of it. This orientation to others, who may be present or absent, is not, for the most part, something which is emotionally flat or neutral. There may be a desire to please another, or to conform to some sense of obligation to that other, or avoid causing displeasure to that other. Certain areas here, those to do with care or feeding for example, may seem to be especially emotionally charged but few issues within family practices lack any emotional component.

We may argue, therefore, that emotions are an essential component of everyday family living and routines just as much as they are, more obviously, involved in family disputes and break-ups. We cannot avoid the emotional complexities that arise when, for example, a wife discovers that her husband has been having a series of affairs, but we may easily pass over the emotions involved in a decision to take the children for a walk in the park. So emotions are not incidental to family practices. But it would be wrong to assume that emotions are all that family life is about just as it would be wrong to assume that places of employment are sites of complete calculative rationality. Sociologists, and other social scientists, have played an important part in bringing to the fore the political economy of family living and it is important to hang on to these more materially based understandings and insights. What we need to think about, in relation to family practices, are ways of linking the emotional to other dimensions of family living. I shall begin, in this section, to look at a couple of ways of incorporating emotions into the analysis of family practices. In the next two sections I concentrate on emotional labour or emotional work within families. I then consider a more recent approach of 'mapping' emotions before asking what is distinctive about emotions in a family context. The concluding discussion will return to considering the possibilities of a sociological analysis of emotions in a family context and asking about the contribution of a practices approach to this analysis.

One, relatively early, way of linking the emotional to other dimensions of social life is to be found in Connell's *Gender and Power* (1987). While, Connell was dealing with gender relations here, the analysis presented could be relatively easily adapted to family living especially given the overlap between family practices and gender practices. Connell focuses on three inter-related elements of 'any gender order or gender regime' (Connell, 1987: p. 99). The first is 'Labour', focussing here in particular on the sexual division of labour. Within family relationships we would wish to extend this to divisions of labour between adults and children. The second is 'Power', and again we would wish to extend this beyond gendered relations to include, as was the case with original discussions of 'patriarchy', the power relations between generations. Finally, Connell refers to 'Cathexis', a term which combines the erotic and emotional dimensions of gendered (and family) relationships, 'relationships organized around one person's emotional attachment to another' (ibid.: pp. 111–12).

The value of this framework, lies not so much in elucidating these elements individually but in arguing that they should be taken together, interacting with each other, in producing a particular gender order or regime. Labour, Power and Cathexis can readily be seen as interacting with each other within, for example, heterosexual marriages in modern societies. These interconnections may be reinforcing or they may produce cracks and contradictions within the system leading to, at the individual level, divorce or separation, or, at a societal level, pressures for change in the laws relating to marriage and divorce. An example of a direct link between these elements can be seen in the idea of emotional power within, say, marriage (Dallos & Dallos, 1997: pp. 70–3). Here power can be seen in terms of imbalances in terms of love, needs and the capacity to meet needs. These needs and emotional imbalances may be realised in everyday domestic activities and divisions of labour.

In an attempt to do something similar, but more explicitly focussed on the family, I distinguished between three 'economies of family living' (Morgan, 2001). These were:

1. 'The political economy', dealing with 'the allocation of resources (including time and effort) within the household and within the wider society';
2. 'The moral economy', the 'ways in which family members reflect upon and account for the decisions that they have made in the course of day-to-day family living'.

3. 'The emotional economy' 'to do with the part played by feelings and emotions within family living' (Morgan, 2001: pp. 232– 3).

Looking back at this, I might now have some reservations about the use of the word 'economy' and the way in which it might conform to an over-rationalistic model even where it gives recognition to the role of emotions. On reflection, it might be more useful to see both the political and the moral economy as having emotional dimensions built into them rather than to postulate a separate emotional economy. Thus a father deciding to take a child for a walk in the park reflects a decision based upon some balancing of pains and pleasures (if I do this, I cannot do something else) and moral obligations. Further, conformity to or departing from perceived obligations also has a built-in emotional dimension. I might also have reservations about my apparently collapsing the economic and the political in the one category 'the political economy', thus merging Connell's dimensions of Labour and Power.

Nevertheless I see this as an attempt to incorporate emotions within any account of family living and to argue that what is important is the way in which it enters into and informs the other dimensions or, in my term, 'economies'. As with Connell's analysis, to remove emotions (or 'Cathexis') would be to limit, severely, the analysis as a whole. But perhaps the most influential attempt to bring emotions to the fore within family and intimate living has been the idea of 'emotional labour'.

Emotional labour/emotional work

It is important to remember that the term 'emotional labour' was originally used in relation to forms of paid employment and to the way in which certain working contexts demand that their employees control or manage their own emotions as well as the emotions of the clients or customers they are serving (Hochschild, 1983). While certain forms of employment (airline stewards, nursing, social work) seem to make particular demands on their employees in this respect, it could be said that an element of emotional labour is present in all working contexts. Most employees, at some stage in their working lives, face disappointments, threats, conflicts of loyalties and so on which all require some managing or control of emotions both on the part of the recipients and relevant others.

There is some fuzziness in the use of the terms 'emotional labour' and 'emotional work' (or 'emotion work') although generally it can be argued that the former term applies to public, employment situations whereas the latter tends to refer to more private or intimate situations. Thus emotional work can take place between lovers, friends or family members. This is the usage adopted by Duncombe and Marsden (1998) who demonstrate how family members are frequently involved in the control of their own emotions while managing the emotions of others. For example, a child encounters some distressing situation such as bullying, a fall, or a barking dog. The mother may feel distress herself (perhaps recalling similar events in her own childhood) but controls her own feelings in order to manage the child's emotions. The use of the term 'emotion work' in this context is designed to give some depth to earlier discussions of domestic labour which frequently focus on measurable and readily identifiable household chores. While these questions of 'who does what?' and 'for how long?' are clearly important they sometimes miss the importance of these more emotional, and less quantifiable, routine interactions within families. Emotional work, it may be argued, is built into everyday family practices.

This 'built in' character of emotional work in the home is one of the ways in which it might be distinguished from the more public forms of 'emotional labour' within workplaces or in encounters between employees and clients. Another is the degree of performativity (Hochschild (1983) refers to surface acting) involved in some of this emotional labour. When the waiter asks 'how are you today?' as you sit at your table you understand this as a performance and, usually, accept it on these terms. While such performances are not absent within domestic life, they may be the occasion of adverse comment or ridicule. This was demonstrated in Garfinkel's experiments where a young person was invited to act as a guest and a relative stranger within the parental home (Garfinkel, 1967).

Emotional labour or emotional work is usually seen as being highly gendered in that it tends to be associated with the work of women or occupations where women predominate. This is clearly in line with widely held gendered models which construct women as being both more emotional and as, therefore, specialists in the handling of emotions. This represents both the more qualitative aspect of the domestic division of labour and a statement about the different characters of women and men more generally. This is the contrast, in Duncombe and Marsden's terms between 'Stepford Wives' and 'Hollow Men' (1998). On the one hand we have wives who are wholly absorbed in their routine emotional work (no longer surface but 'deep' acting) such that it becomes part of their very identity. On the other, we have husbands who find it difficult to express their emotions or to talk about them or find it only possible to express them in forms of activity such as physical sex or violence. Duncombe and Marsden are unhappy about the gender stereotyping that enters into some discussions of emotional work in the home and ask for a more finely nuanced understanding. Men can and do perform emotional work. They are also wary of some of the assumptions about authentic selves which are, allegedly, being suppressed within gendered emotional labour.

Nevertheless the development of the ideas of emotional work in relation to family life has made a contribution to our understanding of family practices. In the first place it shows what is distinctive about much of the activity that is carried out within domestic relationships. It is clearly emotional in that it brings forth emotions in those performing the activities and that it frequently involves managing the emotions of oneself and others. At the same time it has some characteristics associated with work in that it involves the expenditure of effort in directions which are, or seem to be, necessarily required. Further, individuals may feel 'emotionally drained' after dealing with even some of the most routine family practices; getting the children ready for school, for example. The exploration of the relationships between 'caring for' and 'caring about' clearly brings out the ways in which emotions and work interact and cannot readily be separated.

However, it might be argued that the use of the term 'emotional work' diminishes the impact and the significance of emotions when considering family life (Craib, 1995). It might be argued that sociologists are relatively uncomfortable when it comes to dealing with emotions and that the addition of terms like 'work' or 'labour' returns the researcher to more familiar territory. Emotional aspects of family life are present whether individuals are actively engaged in emotional work or not. This leads us to ask what is distinctive about emotional work when performed within the context of family relationships. Moreover we need to ask whether it is possible, sociologically, to talk about emotions within family practices without the addition of the terms 'work' or 'labour'.

Families and emotional work

It will be recalled that the discussions of emotional labour and emotional work had their origin in the analysis of particular working contexts. While particular contexts, especially those associated with the growing service sector of the economy, seem to be particularly identified with emotional labour, other discussions have identified the emotional content of any working environments. And it would not take much elaboration to extend this analysis to leisure activities (consider sport or musical events) or consumption. The relations between work, power and cathexis (to return to Connell's terminology) can be seen to have a very general applicability, especially when applied to gendered systems.

If we are to explore emotions within family practices we need to ask whether there is any special significance in the bracketing of the terms 'family' and 'emotions', especially when almost any human interaction can be described as having an emotional content. Common sense would seem to suggest that this special significance is present and, indeed, obvious. Emotions, it might be said, are built into family life. There may be cultural and historical differences in the ways in which these emotions are expressed or controlled (at times of bereavement, for example) and individuals may routinely and often unconsciously draw upon what Hochschild describes as 'a collectively shared emotional dictionary' (Hochschild, 1998: p. 6). Nevertheless, there is, to say the very least, a strong cultural presupposition that emotions are an integral part of family living. A large part of our literature bears witness to this.

Indeed, it may be said that this previous paragraph is too guarded. Not only, it might be argued, are emotions built into family life but it is also the case that this emotional nexus is the basis of emotional lives more generally. To provide just one example, Roper argues that many of the relationships between senior and junior managers reproduce and reflect earlier relationships between fathers and sons (Roper, 1994). Clearly, the argument at this point would seem to be moving in the direction of a more psycho-analytical interpretation of family life.

There is much to be said for drawing upon the contributions of the various strands within the psycho-analytical tradition. There are the ideas associated with repression and the unconscious and their origin in deeply embedded family dramas. There are themes to do with the long period of dependency on the part of the human infant. There are the ways in which family relationships provide the foundations for the construction of a gendered and sexualised self. I do not wish to dismiss these and other similar insights and, indeed, it would be difficult to prevent these ideas from influencing my analysis at some points. However, I recognise that many of these themes remain controversial and I do not feel particularly well qualified to adjudicate between different psychoanalytical schools. More positively, I feel that sociology does have its own contribution to understanding family emotions (different from but not necessarily better than) and that it is valuable to see how far one might go with these.

To illustrate the ways in which emotions might be handled differently in family and non-family settings, we may look at the contrasting experiences of young children in school and at home (Mayall, 1998). Mayall makes the obvious point that routine childhood ailments or cuts and bruises are dealt with differently in the two contexts:

> Children's accounts of episodes of illness at home endorse the picture of home as a holistic health-care arena ... At home the care they get comprises attention to their physical and emotional condition, in one package.
>
> (Mayall, 1998: p. 148)

At school, in contrast, the care may not go much beyond the proverbial piece of sticking plaster.

Accounting for the differences between 'formal organisations' such as schools and 'primary groups' such as families (although the terminology may vary) has been at the heart of much sociological analysis of modern societies. The point is not around the presence or absence of emotions but more to do with the ways in which and the extent to which emotions are managed and, indeed, expressed.

A key reason why we should suppose a close relationship between family practices and emotions exists is, as we have seen, that members routinely share social and physical space for, often considerable periods of time. Further, again often for long periods of time, some members are entrusted with the care and socialisation of younger members and that these responsibilities (the licences and the mandates) may sometimes involve the denial or frustration of others' wishes. Yet again, the sharing of time and space often has an element of compulsion about it so that exit is an unlikely option for some, especially children and women.

We can multiply these reasons, some of which overlap with more psychoanalytical accounts. The main conclusion must be that the interrelations (explored in previous chapters) between time, space, embodiment and family practices produces consequences which are, in a sense, over-determined. Any one of these elements might be given causal efficacy but it is the interconnections between them that is significant. Out of these relationships we produce the domestic habitus which normalises and routinises these influences and experiences. What is perhaps missing in some accounts of the habitus is the emotional content and the way that it produces frustrations and ambivalences as well as support and security.

An important consequence of these interweavings of domestic time, space and embodiment is to do with the importance of memory, often carried over and transmuted through the years. These memories, pleas-urable, painful or ambivalent, are given shape in photographs and fam-ily albums, in family stories and myths and in repeated conversations. Jokes and repeated phrases and utterances reproduce and simplify these memories as they are exchanged at family gatherings. They enter into processes of family display (Finch, 2007), statements to the effect that we are a family of this kind and that we function as a family.

Illustrations of this everyday 'memory work' is provided by a long-running section within *The Guardian* newspaper's Saturday editions. This is simply called *Family* and regular small items within this supple-ment are to do with food, photographs and music. Thus, for Saturday 12th July, 2008, the section titled *We love to eat* recalls 'Meggies sweet pastry. Meggie, interestingly enough in this account, was a home help and a babysitter, drafted in to enable the author's mother to run a small shop. In common with many of the recipes included in these items, the particular item is easy to make (hence contrasting with the recipes provided by the more professional cooks elsewhere in the Saturday edi-tion), sweet and probably not very healthy. The relationships between food, memories, emotions and family practices remain very strong in many fictional and autobiographical accounts.

Turning to the other sections we have *Snapshot* where another reader has a photograph recalling a little brother's first day at primary school. The photograph (showing a very happy child) clearly deals with a time of mixed emotions for the various participants but the event is used to elaborate an account of how children in London in 1979 were relatively free to roam so long as they were home in time for tea.

The third section, *Playlist*, has a third contributor recalling the song, *In a Little Spanish Town* and also remembering family parties and singing round the piano. Music, particular songs or pieces, is particularly impor-tant in anchoring memories and recalling or re-constructing emotions. One of Tia DeNora's respondents, for example, recalls stopping the car and being in floods of tears on hearing the Brahms Double Concerto on the radio. Her father had recently died and she was recalling how important that piece of music was to him (DeNora, 2000; p. 63).

These examples, repeated on a weekly basis in this particular news-paper supplement, show the importance of memory in constructing and reconstructing family practices. They also show the importance of particular points of anchorage – food, photographs and music – in readily linking emotions and practices in the past with their everyday reworkings in the present. Further, the fact that memories of this kind are popularly represented in newspapers, radio, film and television points to relationships both between individual memory and notions of collective memory and also shows the socially constructed aspect of the close associations between family and memory. As Smart (who includes 'memory' as one of her overlapping core concepts in the analysis of personal life) writes:

> Thus the more work we do in Western cultures on family memories and tracing lines of heritage, the more we contribute to the increasing iconic status of families in our cultural imaginary.
>
> (Smart, 2007: p. 39)

Family rituals and ceremonies provide important occasions for the working and the re-working of these memories as well as for the generation of the emotions associated with family living and family display. Durkheim's discussion of ritual (Durkheim, 1965 (1917); Hochschild, 1998) draws attention to the heightened emotional intensity associated with these repeated practices and the ways in which members feel renewed, through the reaffirmation of what is shared, as a consequence. What perhaps is missing from this account is a more nuanced account of the complexity and the ambivalences of the emotions involved.

We need to distinguish, perhaps, between those family rituals which are associated with major life events (weddings, naming ceremonies, funerals) which frequently involve people from outside an immediate family circle and those other more informal recognition of birthdays or regular family get-togethers. The former are frequently more emotionally charged partly because they are concerned with serious matters, partly because so much (materially and emotionally) is invested in them and partly because the element of display is frequently more obvious. Such events have their own intensity deriving in part from the sense of risk or danger that lies beneath the surface. The popular warnings to the best man in a wedding about not forgetting the ring provide an illustration of the anxieties and risk generated by such events. It is doubtful whether many rings have been forgotten on these occasions but the repetition of this fear highlights wider anxieties about the dangers pervading these semi-public events. At the same time these risks, the dangers that were overcome or by-passed and the parts that did gone wrong themselves enter into the family legacies and provide the material for future stories.

These life-course ceremonies highlight another reason why emotions are built into family living. Families are constituted, dissolved and reconstituted through key events in human life. Formal descriptions of family practices within a given society frequently begin with the demographics, the statistics of births, marriages and deaths. In addition, other matters of key concern such as health and illness, aging and sexuality, are woven into everyday family living. Routine conversations reaffirm that this is what families are really about and these matters are rarely the occasion of emotional indifference. Life-course rituals, therefore, provide a linkage between these key events and concerns in the lives of any individual and ideas of family.

But, as we have suggested, these are not the only rituals associated with family living. These other events may be associated with 'ordinary' birthdays (i.e. ones not associated with the passing of decades or some other landmarks) and anniversaries or they may be simply family get-togethers with varying degrees of frequency or regularity. These events are less public and not, usually, so 'risky' as the life-course ceremonies and may be, as a consequence, less emotionally charged. But they may have their own significance as the occasion for the retelling or the reworking of family stories, the showing of family photographs or video recordings.

I have argued that there is an over-determination, a process of convergence of different streams and influences, which underlies the emotional content of family practices, ranging from the most routine to the most exceptional. This is not to say that family life is always a continuous soap opera, punctuated by frequent and dramatic explosions of emotion. This may be one way of 'doing family' but is probably not the most frequent. For one thing, the emotions we are concerned with are not simply the more dramatic outbursts of anger or affection. What we are frequently concerned with are the minor irritations or the glows of affection that rarely register on the radar screen. For another, as discussed in talking about emotional work, we are frequently talking about the control and management of emotions as well as their expression. This may be a question of being attuned to the tensions that might possibly arise and the adoption of strategies that may by-pass potentially risky situations. Such emotional labour, as has often been noted, is frequently gendered and this itself provides yet another layer to the complexities of everyday family practices. Emotional work within the family is frequently gender work and gender is itself bound up with issues of identity and being in the world.

Mapping emotions

It might be argued, on the basis of what has gone before that families represent particular sites where emotions are not only expected but, to some degree, legitimated. This is partially implied in the contrast between the public and the private. Some degree of control of emotions, both positive and negative, is expected in the public sphere whereas feelings are not only expressed but expected to be expressed in more private or intimate contexts. To some extent this distinction has become blurred with some public concern about intimacies between parents and children on the one hand and the, it is alleged, greater degree of legitimacy attached to public displays of feelings following the death of Princess Diana.

But to argue that family space provides a legitimate or expected site for the expression of emotions does not mean that there are not considerable variations according to, say, gender, social class, generation and ethnicity. Furthermore there may be considerable variation within a domestic residence and this possibility has been recognised in the development of 'emotional maps' as a research tool (Gabb, 2008).

This is a technique associated with the in-depth study of family practices outlined by Jacqui Gabb in her book *Researching Intimacy in Families* (2008). This is a technique that can be used with both adults and children and involves, first the drawing up of a floor plan of the residence and, second, the placing of different coloured emoticons on this map. The different colours represent different family members and the emoticons represent happy, sad, cross and affectionate. The resulting emotional map is then used by the researchers to discuss further aspects of family life and feelings.

For example, mealtimes may be associated with both positive and negative feelings and therefore the appropriate emoticons may be placed around the kitchen table. If mealtimes are generally seen as happy occasions, then the emoticons will present this; if mealtimes are often tense or fraught occasions the representations may depict anger or unhappiness. Similarly entrances and exits (say to particular bedrooms or the front door) may be seen as areas of conflict and recognised accordingly. An extended example may be taken from Gabb:

> Kelly's map revealed her creation of private space and a distinct individual identity. The PC, located in the office, afforded her the chance to be (virtually) with her friends, away from family. Her bedroom is a space for her and her friends/boyfriend. The entrance (bedroom door) is identified as a place where angry exchanges occur as parents intrude on her physical and emotional territory.
>
> (Gabb, 2008: p. 178)

In principle the emotional mapping could extend beyond the household and, for example, include locations in the neighbourhood where individuals feel free or threatened. Further, the emotional range could be broadened to include more emoticons allowing for more complex feelings. Nevertheless the emotional map serves as a useful way to concentrate thinking about the place of emotions within the household. We may see some of the complexities involved where, for example, two different family members put different emoticons in the same spot. We can see that emotions are relational and emerge in family encounters. We can see that, although the term 'map' has primarily spatial connotations it is also closely associated with time; mealtimes or bedtimes, for example. Indeed we can see the close relationships between time, space, embodiment and emotions. Finally memories of family life may focus not simply on particular residences but on places within or beyond those residences where one felt threatened, unhappy, happy or free.

Families in context

I have argued that there is a danger in restricting a discussion of emotions to accounts of emotional work an approach which, while valuable, runs the risk of playing down the centrality of emotions within family practices. In principle, almost any family practices, because they are carried out by and between people who are and have been related to each other, may be emotionally charged. We may think especially of feeding and providing and caring and protecting but also of everyday excursions, watching television or planning holidays. Feeling comfortable or uncomfortable in another's presence is as much to do with emotions as explosions of anger or strong affection.

The question remains as to the extent to which the family is uniquely involved in these linkages between practices and emotions. I have argued that even if we put to one side more psychoanalytical interpretations there are good reasons for us to suppose a strong connection. But family life is not unique in this respect. Neighbourhoods, networks of friends and working environments may generate something of the interconnections between time, space, embodiment and memory that seems to be characteristic of much family living. The musical memories referred to in DeNora's study are not confined to family relationships (DeNora, 2000). A strong competing environment would be schooling where, similarly, we have a concentration of time and space and embodiment, the overlapping of generations and matters strongly connected with a growing self and a developing identity. Seeking to renew contacts with former school friends is almost as popular an activity as tracing family connections and memories of school, positive and negative, may be as intense as family memories.

What this points to is that there are strong overlaps between sets of family relationships and other sets of relationships in this and in other respects. In the case of schooling this may, in modern societies, be particularly obvious but this is partly due to the coupling of home and school in practice as well as in theory. Discussions of the different and overlapping responsibilities of parents and teachers – in matters to do with school attendance or completion of school assignments, for example – continue and reflect they way in which the two sets of practices are interwoven. The availability or otherwise of 'good' schooling may be a powerful influence on family practices when it comes to deciding where to live. But even without these state-mediated relations between home and school it is likely that the emotional recollections of school may merge with, or even be more intense than, other more family-based experiences.

We may say therefore that while there are strong reasons for the emotional character of family practices these do not in themselves argue for the absolute uniqueness of family life. Sociologists are becoming increasingly aware of the role of emotions within formal, more rationally based, organisations and while there may be continuities between families and other spheres of life in this respect it is likely that these other arenas are also distinctively significant in the construction and control of emotions. In other words, a foreman may be disliked because of his position in the workplace and his personality and not simply because he reminds someone of his father. This is not to say that emotions are uniformly spread across the whole spectrum of social life and it is likely that emotions have their own logics in different contexts. Just as we can draw emotional maps in relation to family life, so too may we be able to provide similar mappings for all the areas of social life in which we find ourselves.

Concluding remarks

I hope to have shown in this chapter that emotions are a key aspect of family living with the implication that any account of family that excludes emotions will be defective. At the same time, this raises questions for the possibilities of a sociological analysis of an area of life which appears to belong to psychology or more imaginative, literary accounts. The analysis of emotional labour or emotion work takes us some way in this direction but some readers might feel that the use of terms such as 'labour' or 'work' has the effect of holding emotions at a distance. Put another way it might be asked whether the analysis of emotion work is all that can be said about emotions and family practices.

To illustrate the point consider the experience, briefly noted in an earlier section, of a mother taking a child to school for the first time. This is often seen as an important transition in family life and one that is accompanied by, often ambivalent, emotions. 'Emotion work' may be part of this experience – in talking with the father or with other mothers for example – but it is doubtful whether it is the whole story. An observer may infer that the mother is doing emotion work in holding back her tears or fears but this takes the analysis to a different, more external, level. It would seem that the idea of emotion work can provide some useful insights to what is going on within family practices but does not tell us the whole story.

We are clearly at one of the points where sociological analysis overlaps with other modes of enquiry and while this should encourage some measure of caution or tact on the part of the sociologist (or other specialist) it should not inhibit the analysis if it be proceeding in this direction. In the course of the brief discussion in this chapter we have seen several suggestions of how the sociological analysis of emotions and family practices might be conducted. One of these is the analysis of 'emotional mapping' an ethnographic tool which brings together discussions about gender and generation, the spatial dimensions of family living and everyday family practices. This is an approach which is sensitive to what people say and how they say it and which does not reduce the analysis of emotions to some of the key variables of sociological enquiry.

Another more sociological set of questions concerns the positioning of emotions with family relations. Families are not only seen to be legitimate sites for the expression of emotions but, in some senses, required locations for emotions. This contrasts with some perceptions of the public arenas of work and politics where emotions, while recognised, are still seen as competing with and probably subordinate to more rational considerations. As I have argued earlier the close mapping of private/public on to emotional/rational may have declined but the broad picture remains much the same. Tears may be permitted in public life but only, one suspects, on certain occasions and with limited frequency.

To explore the close association between families, and other intimate relations, and the legitimate expression of emotions requires some detailed historical analysis in terms of the overlaps and interconnections between the public/private divisions, rationality and feelings and gender, to name only some of the key divisions. Thus we may explore the development of the idea of a human relationship in the strong sense of the word so that it does not simply refer to a connection between two individuals (as in a kinship diagram) but a deep, complex and multilayered set of linked strands that are seen as being of particular importance in a person's life. We may explore the idea of intimacy with a particular focus on confiding and the sharing of emotions (Jamieson, 1998). We may explore the role of a whole host of professionals, advice experts and television presenters in developing the sense that not only is it legitimate to express emotions within a family context but also required to do so. Put another way, apparent failure to express emotions within family and intimate contexts may be a sign that something is wrong with a particular set of relationships (Clark & Morgan, 1992; Morgan, 1985).

These questions, demanding the analysis of broad cultural variations and historical shifts in structures of feelings may be somewhat remote from family practices. However we may begin to explore how these themes are played out in everyday life through the analysis of both emotional mapping (already discussed) and emotional talk. To consider the latter this is an extract from a much longer discussion of the idea of home from an interview with a young Liverpool woman:

> Erm, stability, I think erm plays a big part in what I need to feel secure. Erm, warm, erm, cosy, er, decorated in colours that I like having around me. And having me little erm, sentimental little bits and pieces. Me cats!

Clare Holdsworth and myself found that the words 'stability', 'warm', 'cosy' and so on occurred frequently in respondents accounts of the meaning of home Holdsworth & Morgan, 2005). We can also see, in this quotation, the association between the idea of home and a sense of self and personal possessions. These particular images of home were to be found among responses from Bilbao and Trondheim as well as from Liverpool and were to be found in responses from individuals who had not had particularly positive experiences of 'real' home. It can be seen that these words have a strong emotional content but refer as much to public discourses as they do to individual experiences.

Or consider another aspect of emotion talk. When people clearly show sad emotions through a change in voice or through tears they frequently apologise to the listener(s) for this apparent breach in what is normally expected. In part the felt need to provide such an apology supports writers such as Goffman who point out the ways in which individuals not only present themselves but also work at preserving an ongoing interaction (Goffman, 1967). The speaker is embarrassed at causing (it is supposed) embarrassment to the hearer. The interaction is maintained through the apology and, in return, the proffered tissue or expression of concern. We may attend to these micro, but often highly significant, interactional issues or we may ask questions about the nature and significance of tears in society (Carmichael, 1991). Discourse analysis around emotions may point us to individuals and their social relationships or to the wider context in which emotions are expressed and managed.

I now return to the remaining question about the ways in which the practices approach may contribute to these discussions about emotions in family and intimate relations. Many of these points have been raised earlier so a detailed recapitulation is probably not required. The first point is that the practices approach emphasises 'doing'. This has clear affinities with the emotional labour or emotional work approaches and while I have suggested that there may be limitations to these approaches they clearly have a place in the analysis. At the same time it may be suggested that family practices as a whole are broader than those practices within family constellations which might be identified as 'emotional work'.

Yet, second, the fact that a particular family practice might not be identified as emotional work does not mean that it is devoid of emotional content. The emphasis, within the practices approach, on the everyday and the repeated, also points to emotions which barely register on the interpersonal Richter scale. The quiet pleasures of the familiar or the mild annoyances associated with the habitual are nothing to write home about but are frequently what home is about. The practices approach, with its emphasis on the routine and everyday takes our understanding of emotions beyond anger and distress.

Finally, I return to the reflexive character of family practices. This is the fact that family practices are orientated to others who are defined as or in relation to family members and that, in enacting these practices, we are affirming or reaffirming particular sets of family ties. I have argued that this is rarely a matter of indifference. Whether the others, who are taken into account in the performance of family practices, are there out of a sense of duty or obligation or through deep liking and personal preference, emotions are involved. Further, since we are talking in relational terms we can see how emotional practices are not simply matters for an individual's 'inner' life but are shared with others within family, and other, configurations.

8
The Ethical Turn in Family Studies

Introduction

Some readers might find the title of this chapter slightly strange. It seems to suggest that a linking of family practices with ethical concerns is something of recent origin whereas it might be supposed that morality has rarely been far away from family living. Repeatedly, it might be argued, politicians, religious leaders and moral entrepreneurs of all kinds have made connections between family and morality. Measures of family breakdown may be taken to be indices of a wider moral decline and, equally, a weakening of family ties may be seen as contributing to a wider social and moral breakdown. So what is new?

The quotation following this paragraph might convey the essence of what I want to discuss in this chapter. It is taken from interviews that formed the basis of a comparative study of young people leaving the parental home (Holdsworth & Morgan, 2005) and here a mother is talking about the fact that her daughter's boyfriend regularly stays overnight:

> Erm, I, yeah, because, I just feel, she's nineteen. And erm, she'as an adult, isn't she, and its, her, well I don't know if, I don't know if, sometimes I don't feel its right because of Connie and Paul [younger siblings] knowing he stays. Erm, but they just accept it, but erm...I don't know, I just feel that she was quite sensible, erm, and at the end of the day I thought, it's if they're not here, where are they? I don't know, Jackie [interviewer], if I've done the right thing. (Interview transcript).

What is clear from this quotation is that the interviewee is conducting some kind of ethical debate with herself and with the interviewer. There is a lot of hesitation and an attempt to introduce and to weigh up all the relevant considerations; age and ideas of adulthood, impact on other family members, the limits of parental surveillance and so on. Almost certainly, this is not the first time that the subject has expressed (if only within herself) these considerations. She is dealing with a very everyday issue, one faced by most parents in similar social and life-course situations. And, while there are no references to abstract religious or moral precepts, she is very much concerned with doing the right thing and with presenting herself as a moral individual.

These issues are at the core of the concerns of this chapter. We are concerned with everyday, practical ethics rather than the more abstract discussions of moral philosophy or religious doctrine. Most of all, we are not concerned with 'moralism' or with elaborating moralising discourses. Instead we are concerned with 'moral reflections on social practices' (Sevenhuijsen, 1998: p. 37). Later, I shall look at some of the wider discourse but through the lens of these more everyday practical activities and reflections.

These concerns have come more to the fore in family studies within the last 15 years or so. In part, they have arisen out of detailed investigations into everyday family practices and dilemmas to do with, say, the provision of informal care, inheritance or the management of family relationships following divorce or separation. These research projects and these wider debates have been influenced by feminist writings which have attempted to place more immediate concerns about the provision of care and the practices of women within families within some wider philosophical context. Thus Sevenhuijsen (1999) argues that a feminist ethic of care focuses on the 'self-in-relationships', on 'situated questions of responsibility and agency' (p. 9) and something which is aimed at 'concrete needs in concrete situations' (37). All this can be linked to growing discussions of relationality in personal life (Smart, 2007). What we have seen, therefore, has been a set of converging research projects and scholarly debates which have argued for taking issues of ethics seriously but in a way that might be different from earlier moralising discourses (see also, Sayer, 2005).

There are some links in this chapter with what has gone before. The most obvious connections are with emotions, the theme of the previous chapter. When Adam Smith (1759[1976]) wrote about 'moral sentiments' he was stressing the intimate connections between the ways in which we feel about other people and ourselves with questions of morality. It is essentially a relational approach. From a quite different, although converging, perspective, Nussbaum also sees emotions in relational terms and as involving judgements of value (Nussbaum, 2001). To place a value on something is, among other things, to express an emotional response to what is valued. This close relationship between emotions and morality becomes especially apparent when looking at the everyday ethics which constitute the core of this chapter.

There are also links with issues of the body and embodiment. Ethical questions, how we relate to other people, do not deal with these others as abstract spirits or unspecified others but as embodied and feeling individuals. The concerns felt by the mother in the quotation above are to do with embodied practices and with embodied individuals located in a particular space. Issues of care, sexuality and personal space all, as we have seen, are embodied issues and all are frequently the subject of everyday ethical debates and decisions.

It should not be too difficult to find links to others' chapters within this book but the overall point is to underline the argument that ethics are not a separate or rarified area of discourse but are closely bound up with everyday practices. It is here, perhaps, that we need to make a distinction between ethics and morality. The two terms have been used more or less interchangeably up to now and this probably reflects some everyday usages. However, I shall from now on use the word 'ethics' to refer to the reflections and debates that surround everyday family-based practices and dilemmas. I see the word in more practical and situational terms. I reserve the term 'morality' to refer to the more public discourses about right and wrong and how these might apply to family living. In practice, as we shall see, this variation on the private/public distinction might not always work but it should serve as a rough and ready point of departure.

Related concepts

The Moral Economy

The term 'moral economy' has been used by a variety of scholars in different disciplines in order to highlight the fact that, even within a dominant market economy, there are also shared norms relating to such themes as reciprocity, obligations, fairness, custom and interdependence. Contrasts are made between the political or the market economy and the moral economy although there may be some discussion about the extent and the ways in which the two economies articulate with each other or the extent to which the moral economy represents an older, pre-capitalist, social order which was largely replaced by a market economy. Other contrasts might be those between exchange value and use value (with the moral economy being identified with the latter) or between individualistic and more relational models of economic life (Minkler & Estes, 1991).

An example of an area where the idea of the moral economy has been found to be useful is in relation to issues to do with aging and the life-course (Minkler & Estes, 1991). Here we can see a whole series of questions arising in a context where emphasis is placed upon employment and thus, in many cases, on able-bodied adults. Questions include relations and obligations between generations, between dependence and independence and between the able-bodied and the less able-bodied. Many of these issues have a particular resonance in the context of family relationships.

Why is the term 'economy' used in this context? In part, this usage represents a challenge to the idea that the only model of economic relationships is the one represented by the market economy. Yet we are continuing to deal with various claims and counter claims and with the allocations of resources and services and these are all matters which might be understood to be at the heart of any economic system. At the level of everyday ethics the question is often not one of conformity to normative expectations but of the balancing of different sets of obligations and claims from different significant others (Finch & Mason, 1993).

It is important to stress that these everyday questions arise most frequently (although not exclusively) within family relationships and are intimately bound up with everyday family practices. It would be overstating the case to argue that some of the core concerns of the moral economy – fairness, obligation and reciprocity for examples – have their origin in family relationships but it is certainly possible to identify a contrast, and sometimes a tension, between the moral economy of family practices and the more abstract rationalities of a market economy.

The generalised other

The idea of the moral economy looks at the ways in which moral concerns are embedded in the wider structure of economic and social practices. Here we are particularly interested in the ways in which these concerns are played out at the level of family and intimate practices. While issues to do with obligations and conformity to collective expectations are involved in the moral economy, the idea of the 'generalised other' deals more directly with these (Holdsworth & Morgan, 2007). In Mead's account (1962[1934]), the generalised other is an extension of the basic human interaction of taking the role of the other into account when engaging in particular practices. He makes a distinction between 'play' (the actual activities and practices of players in the course of a particular game) and the 'game' which

> involve[s] something which is more than the sum of individual co-players, such as notions of the rules of the game, ideas of our team or our side or wider notions of sportsmanship.
>
> (Holdsworth & Morgan, 2007: p. 403)

The generalised other can become part of the internal conversation that individuals have with themselves in considering the right thing to do or, retrospectively, the moral implications of their actions. The quotation earlier is a good illustration of this kind of debate, part internal and part external.

Mead's account has sometimes been characterised as representing some kind of 'group mind' although this is probably a misreading. In any event, it is probably best to shift the analysis somewhat to explore the actual processes of generalising and the way in which these enter into everyday moral discussions. Consider this quotation from 'Lynn' who is still living in the parental home at the age of 35:

> It just suits me. I know there's a stigma about people living at home, you know, at my age. When you think that, you know, the likes of me friends are all either married or married with kids, or living with somebody. And I haven't got that.
>
> (Holdsworth & Morgan, 2007: p. 410)

Here Lynn refers, in a general way, to her knowledge that there is 'a stigma about people living at home' at her age. This general, if non-specific, awareness is linked to perceptions of what her friends are doing. This is somewhat more focussed but still at a fairly high level of generality. If we are to talk about 'the generalised other' it is probably in relation to these processes of linking general ideas to particular sets of individuals such as friends or other family members.

The links between ideas of the generalised other and family practices are two fold. In the first place, family practices – family life in general – provide a convenient framework for the conduct of these moral dialogues. There is frequently a ready assumption that the listener will know what the speaker is talking about and will be able to make connections with her own family experiences or practices. Second, identifiable family members (especially, one assumes, parents) will be singled out as a source of some of these more general prescriptions. Everyday utterances and proverbial expressions which outline features of the moral universe are frequently mediated through significant family others. These general precepts may be passed down through generations. Again, family members are not the sole source of these everyday guidelines but they are almost certainly among the most important, even where they might be rejected later in life.

Ethical family practices

The first thing that can be said about ethics in relation to family practices is that we are dealing with everyday concerns. While individuals may deploy general statements in the course of accounting for any particular planned or completed course of action the actual level of concern is with particular family-based issues. It is *this* mother's sickness, *this* brother's debts, *this* couple's infirmities that provide the focus of concern rather than family responsibilities in some more general sense. Moreover, these particular and immediate concerns are embedded in sets of ongoing inter-dependencies and commitments so that obligations in one direction have to be balanced with accumulated obligations from some other source within one's immediate network.

Illustrations are readily available in recent research literature. There are the obligations placed upon or assumed by family members in terms of the provision of informal care for a sick, elderly or disabled relative (Finch & Mason, 1993). There are issues to do with the defining or redefining of significant relationships following divorce and remarriage or on becoming a stepparent (Smart & Neale, 1999; Ribbens McCarthy, Edwards and Gillies, 2003). There are decisions to do with the passing on of family property, heirlooms or 'keepsakes' to others (Finch & Mason, 2000). As Mason notes, 'investing inherited money is not, therefore, governed by an economic rationality, but a relational form of moral reasoning' (Mason, 1999: p. 14). There are issues to do with individual sexualities such as 'coming out' or entering into a gay or lesbian relationship (Weeks, Heaphy & Donovan, 2001).

These concerns (which could be multiplied) although diverse have several things in common. In the first place, although they involve practical arrangements they do not remain at this level. Issues of ethical judgements and concerns about 'the right thing to do' are not far below the surface. Ethics here is not simply a question of the language that is adopted but also one of making choices which are seen as being consequential for oneself and for significant others.

Second, in making these decisions, individuals do not routinely draw upon abstract normative prescriptions as a source of direction. Finch and Mason asked whether there was a clear consensus about responsibilities towards relatives in Britain today and came to the conclusion that there was not. Such responsibilities were not seen as 'automatic or unlimited' (Finch & Mason, 1993). General prescriptions may play a part in the overall process of moral deliberation and accounting but these are woven into the more complex and more everyday ethical practices.

Third, as we have seen, these ethical decisions do not simply involve one family member and another member who is seen, in some way, to demand consideration or attention. Each individual is involved in networks and sets of relationships which have their own historical sets of inter-dependencies and which need to be taken into account when making a particular decision in relation to a particular other family member. These complexities may multiply following certain family transitions such as divorce and re-marriage where these others may come to include ex-partner's parents, new partner's parents and children and so on.

Taking all these features into account – the need to make complex decisions involving a variety of significant others – it can be understood why the term 'negotiation' has entered the sociological literature dealing with these issues. In one sense this term, with its echoes of the public spheres of international or industrial relations, might seem a strange one to use in the context of family life. But the term in fact has a long pedigree in sociological analysis and captures something of the fluid and complex nature of everyday family living where the political, the emotional and the moral economies meet and interact with each other.

Negotiations, whether overt or covert, help us to understand the decisions that individuals make within family networks, especially when we understand the past relational histories of the social actors involved and the outcomes of previous negotiations. The use of the term 'negotiation', however, should not lead us to suppose that models of rationality will be the most appropriate frameworks of analysis. At all points we are looking at the inter-relationships between rationality and everyday ethics with, if you like, moral rationalities.

We may see this in the everyday language which is used to describe and account for these decisions and practices. The mother, quoted at the beginning of this chapter, asks herself, and the interviewer, whether she is doing the right thing. Or consider the everyday phrase, 'it was the least I could do'. This is a phrase that points to open-ended obligations and a desire to reach or go beyond some unspoken minimum of what might be expected. It is also, perhaps, a disavowal of any kind of claims that one might be doing something special, a desire to place what might seem to be especially virtuous or heroic into the realms of the normal and the everyday. Everyday moral reasoning and accounting is full of these complexities which become subsumed in routine phrases.

There are also some more explicit words that have a more direct ethical resonance. One of these is the idea of 'fairness' (Ribbens McCarthy, Edwards & Gillies, 2003: p. 106). This is a good example of a formally abstract term which is used situationally within family practices. What is important here is not simply the actions or decisions taken but the way in which they are seen by those involved in the actions. The use of the term 'fair' also defines the range of others to whom these ideas might be said to apply in any particular situation. This may be some kind of balance between different children (or sets of children in the case of reconstituted families), adults and children or different sets of adults.

Another idea which often appears in family discourse is the idea of 'sacrifice'. In modern Britain, for example, this may be used in relation to attempts to secure the best education for one's children (Devine, 2004). These may include moving house, giving up on some pleasures or luxuries or taking on extra work. These particular moral accounts are bound up with wider notions of 'putting the family' (i.e. the children in most cases) 'first' (Ribbens McCarthy, Edwards & Gillies, 2003: p. 32) or, more generally, 'being there for each other' (ibid.: 36). As one informant stated: 'having children means that you have to give up being selfish'(ibid.: 40). Ideas of family and everyday morality frequently appear easily within the same frame of reference.

Ideas of morality and everyday ethical decisions within the family are not simply conveyed in these ordinary phrases and words such as 'fairness' and 'sacrifice'. They also appear more complexly in moral tales. Family members frequently talk about each other to each other or to non-family in the form of stories or narratives. Telling stories is a key family practice just as stories are woven into many, probably all, areas of social life. Such stories might be very simple, sometimes not more than a few sentences. 'I dropped into Dad's to see if he was alright. He seemed fine but I noticed that the fridge was getting a bit empty'. In Ribbens McCarthy et al.'s words:

> Moral tales position the talker, others within the tale, and the listener, so that certain feelings and courses of action are regarded – explicitly or implicitly – as right and proper.
>
> (Ribbens-McCarthy, Edwards & Gillies, 2000: p. 8)

These stories provide ethical accounts of what was done and, sometimes, what was not done where a particular course of action might have been expected. There are occasions where individuals need to account for behaving in ways that might seem to be uncaring or neglectful for, in Finch & Mason's words, providing 'legitimate excuses' (Finch & Mason, 1993). Whether or not such excuses are heard or accepted as legitimate will, of course, depend upon the particular family constellation and the interweaving histories of those within it.

An important strand, linking the telling of moral tales and the construction of a morally adequate self is the idea of reputation (Finch & Mason, 1993). Reputation, as the word implies, is not something which is established overnight or in relation to one particular action. It is something which is built up over time in relation to sets of significant others. Reputations are not just or simply moral reputations (one can have a reputation as a skilled worker, for example) and they are not confined to family configurations. However, the relatively closed and inter-related character of many family configurations means that moral reputations can be built up over time almost independent of the wishes of the person whose reputation is being constructed. Thus, over time, individuals may be constructed as reliable, mature, responsible or the opposite of these terms.

Moral accounting and moral tales are, as has already been indicated, important in the construction of a self. As Hekman (1995: p. 128) writes: 'our moral beliefs define us as persons'. But this is not just a matter of beliefs; indeed it is more a matter of behaviour, how we account for that behaviour, and the others to whom this behaviour and accounting is directed. What is important is how we are seen and how our actions are weighed and understood and, again, this process of understanding requires prior knowledge and shared histories. What makes people moral agents, Neale and Smart argue, is not a question of making the right decisions 'but whether they reflect upon the decisions they take and weigh up the consequences of their actions' (Smart & Neale, 1999: p. 114). Again, we are not talking about abstract questions of moral philosophy but everyday family practices and how they are presented, accounted for and understood. As Sevenhuijsen states: 'the moral agent in the ethics of care stands with both feet in the real world' (Sevenhuijsen, 1998: p. 59). The activities and the words used to describe or account for these activities might seem to be everyday and even banal but they relate to some very fundamental issues of moral identity and how we are seen in the world.

These strands in exploring ethics and family practices may be illustrated by Duncan and Edwards' many-stranded study of lone mothers (Duncan & Edwards, 1999). They begin by noting how lone motherhood has been the focus for moralising discourses over several generations and in numerous social contexts. However, this is not, in the twenty-first century, a matter of simple condemnation or stigmatisation of a particular grouping of women. The image of the fallen woman has been replaced by a range of moralising discourses conducted by different sets of people within society. Thus lone mothers may be seen as a social threat, as a social problem, as women adopting a particular, and chosen, life style or as women escaping from patriarchy (Duncan & Edwards, 1999: pp. 25–42). Interestingly, although Duncan and Edwards do not make this point, as there has developed a more complex set of representations of lone mothers there have also developed simpler representations of feckless, irresponsible fathers.

These discourses around lone motherhood may be seen as examples of public moralities as opposed to the more everyday ethical practices. Yet, they are related. Lone mothers are clearly aware of, and respond to, these public discourses. Individual mothers, in interview responses, reflect on their own practices within an awareness of these public representations:

> Yes [I think of myself as a single parent]. I've brought [my son] up from a baby. I've got up for him, changed him, fed him. If it wasn't for me I don't know where he'd be. His father's no good.
>
> (Grace – a white working-class lone mother; Duncan & Edwards, 1999: p. 50)

This is clearly more than a simple account of everyday parenting practices; it is a language of moral responsibility in relation to her son and moral positioning in relation to the father. Another respondent provides a clearer illustration of the way in which a mother negotiates with the public images:

> I would say single parent. Lone parent has the implication of lonely and it doesn't sound very positive. It just sounds sad. And single parent is a celebration of being single. I can do it on my own.
>
> (Charlotte – a white 'alternative' lone mother living in Brighton; ibid.)

Duncan and Edwards show, among other things, the ways in which lone mothers do not simply respond to these public discourses but also routinely reflect about the ethical implications of their everyday parenting practices. There are a range of key practical decisions such as, centrally, whether to seek paid employment or not but these decisions are made in terms of 'what is best and morally right for themselves and for their children' (Duncan & Edwards, 1999: p. 109). These decisions are not made in isolation but with reference to others within their social networks and localities.

Throughout the book, the authors use the phrase 'gendered moral rationalities'. This phrase serves as a reminder that there are not two distinct spheres of decision making, the one based upon rational calculation and the other based upon ethical considerations. The two are very closely woven together. Further, they are gendered. This is less a question of the complex debates about the extent to which men and women have different patterns of moral reasoning and more one of recognising that the decisions made and the considerations that are taken into account are centrally bound up with the lone mothers' identities as women. In 'doing' parenthood and in reflecting upon and accounting for their parental practices, these lone mothers are also 'doing gender'.

The 'ethical turn': Possible limitations

These discussions of practical ethics within family practices are important in that they run counter to some public discourses whereby certain family-based identities (lone mothers, for example) or practices are held up for disapproval. Within these wider public discourses, married is better than unmarried, heterosexual is better than gay or lesbian and stability and continuity is better than the reverse. It is remarkable how frequently these public debates around family identities and structures return as a way of marking out political differences. The more sociological understanding of everyday family and relational practices remind us of the continuing importance of 'doing the right thing' and of being seen to be doing the right thing. In other words, having children out of wedlock or going through processes of divorce or separation are not necessarily to be taken as examples of wider currents of selfishness or individualism. Sociological accounts of family life highlight the importance of everyday moralities and practical ethics.

Clearly, then, the 'ethical turn' in family studies has made considerable and long-lasting contributions to our understanding of the family within late modern societies. But, are there limitations to this approach? To ask this question is not simply to make the obvious point that individuals may make the 'wrong' ethical judgements or choices in their everyday family practices. Nor is it to make the point that not all family members present themselves as morally serious individuals all the time. Actions may be carried out habitually or impulsively and with little attempt at moral accounting. The readers, as well as the researchers, may have their own ideas as to the moral adequacy or otherwise of the accounts with which they are presented.

There are two very general points that may be made about the presentation so far. In the first place, the overall framework of reference is within networks of family relationships. As with other areas of discussion, family relations are constituted through these moral decisions and evaluations. This, in many cases, may be a function of the original research project which will almost certainly have some reference to 'family'. Second, the data for most of these accounts comes from interview material, often of a more open-ended kind. Such a technique (often defined as a 'conversation with a purpose') encourages respondents to enter into the kinds of ethical discussions, with all the hesitations and qualifications, that constitute the core of this chapter. In short, the researcher bears some responsibility for constructing this particular form of moral reasoning conducted within a familial context.

But how far do these conversations and the context within which they take place limit the overall range of ethical issues under consideration? As an illustration of a possible wider context consider the following statement from a middle-class parent talking about private education:

> I basically don't think it's fair, but I think at the end of the day Bruce and I always agreed that the children, the family comes first. I suppose that's rather wet really but that's what we feel. If politics interferes, that has to go.
>
> (Quoted in Jordan, Redley & James, 1994: p. 142)

There are several interesting points about this quotation. One is the near conflation of 'family' with 'children'; this is not an uncommon identification. More important for the present purposes is the awareness of a wider framework of reference against which their decision (private education of the children) may be judged. This is seen in references to wider notions of 'fairness' and possibly being seen as 'rather wet'. Again this may be taken as an example of the deployment of the 'generalised other' referred to earlier in this chapter.

The book from which this quotation is selected is called *Putting the Family First* (Jordan, Redley & James, 1994), and the title reminds us that issues of presenting oneself in morally adequate terms do not remain with decisions made within a particular family context. The notion of 'putting the family first' implies some points of comparison outside the family. This may be other families (i.e. putting *our* family first), a community or neighbourhood or, most likely, some more abstract notions of fairness or justice.

These everyday phrases 'construct the nuclear family as one's primary unit of account' (Jordan, Redley & James, 1994: p. 38). Somewhat paradoxically, Jordan et al. link this focus on the family as a key unit with wider notions of individualism in late capitalist Britain. Thus parents make decisions about their children's education, a prime focus of concern when it comes to putting the family first (ibid.: p. 141). These decisions might entail parents making sacrifices in relation to forgoing other projects or in terms of additional expenditure associated with relocation or purchasing private education. This may mean also going against their own political principles or ideas of fairness. It might also mean behaving in ways which might be judged as dishonest when, for example, they claim religious allegiance in order to get a placement in a faith school or falsify information about their address.

In arguing for 'putting the family first', family members are making similar claims for moral adequacy to those justifications about decisions reached within family contexts. However, there does appear to be a difference. In the case of the former, the family appears to be a terminal value in that it is difficult to imagine any context where the family might take second place. Even entities which might be thought to take precedence (such as the state in times of war) might be linked to family values so that in defending the one you are defending the other. This is different from some of the other decisions discussed in the main part of the chapter where decisions are made in terms of this family member as opposed to this other family member or this course of action as opposed to other courses of action.

The intention of this discussion is not necessarily to find the broad drift of the 'ethical turn' within family studies to be inadequate. I hope that I have shown that these studies have been providing illuminating insights into family practices and everyday ethical reasoning. But I hope that I have suggested complexities which may sometimes be ignored when occupied with the immediacies of everyday practices and decisions. These complexities take us to the boundaries of any particular family constellation and raise further questions about the positioning of family relationships as a whole within modern society.

Concluding remarks

I have outlined some of the key features of the 'ethical turn' in family studies, that is of a cluster of studies and writings which emphasis everyday ethics in relation to family practices and where some of the key words are 'moral tales' and 'moral adequacy'. I have argued that, despite some possible limitations, this represents a significant body of work, one which enhances our understanding of everyday ethics as well as of family practices more specifically.

In this conclusion I address two specific problems. The first is the extent to which there is any kind of special relationship between ethics and family practices which makes it almost natural that the two should be discussed together. The second is whether there is anything special in the practices approach, the subject of this volume, which makes a distinct contribution to the sociology of ethics.

In relation to the first set of questions it might be noted that it may be argued that family living represents the source of much, if not all, of our ethical understanding and practices. Ideas of care, especially that of a parent (and especially a mother) for a child may be seen as paradigmatic for wider practices of care beyond families. Notions of sharing, commitment, responsibility and obligation may be seen as a playing out of values and practices that were learned within the home as part of the socialisation process.

An obvious response to these popular arguments is that they represent a highly idealised representation of family life which plays down issues of inequality, neglect and abuse within family relationships. Nevertheless these ideas at a discursive level often remain quite powerful. When politicians argue that a sense of right and wrong depend upon strong family relationships they are not simply pointing to direct parental instructions but more, one might suggest, to the everyday examples that are supposedly provided in routine family living. Learning to get to grips with the complexities of everyday family life, the inevitabilities of disappointment, the necessity to make awkward decisions, the importance of developing some kind of moral reputation, is part of what it means to be brought up within a family context.

There are two comments that may be made about this line of argument linking ethical behaviour to family practices. The one is that it does not make a commitment to any particular family constellation. The families concerned might be relatively closed and highly nucleated or they may be relatively open and weakly bounded, including a wide range of kin and others. The parents may be gay or heterosexual and the family constellation may include the children and relatives of former partners.

The second is that, as I have argued in the previous section, ethical debate and discussion may begin with family practices but does not, and it could be said, should not, end there. When it is argued that a 'good' family provides a firm foundation for ethical behaviour in later life, this argument frequently rules out of consideration questions of gender inequality or our responsibilities to those beyond our immediate family networks. Family practices may provide a good foundation for moral individuals but they may also be the basis for notions of exclusiveness, 'amoral familism' (Banfield, 1958) and some of the distortions that may follow from 'putting the family first'.

I have here only touched upon some complex issues that probably stray beyond the limits of sociological analysis. While it can be argued that, in terms of discourse, it is possible to establish strong links between family and ethical practices, in principle matters are more complicated than that. However, there are other reasons which point to strong, if not necessary or inevitable, linkages between family and ethical practices. This is to do with the relationships between family practices and life events.

I have already touched on this point at several stages and will return to it in the conclusion. This serves a reminder for the fact that family practices exist in time and are, in some sense, about time. If it is true that children are not 'beings' but 'becomings', the same is also true for all identities and relationships within family constellations. Change and development are at the heart of what family life is all about and the key changes revolve around the fundamental life events of birth, coming together, separation, sickness and death. Since much ethical discussion itself, and inevitably one might assume, also addresses these key life events there is no surprise that family life is linked to ethical considerations. As with the previous argument, there is room for further debate about these connections, but there is no doubt that the links are strong. The greatest proportion of individuals attending weddings, birthing ceremonies and funerals will be other family members. Even in the case of celebrities, there may be a distinction between a public event and a more private ceremony for 'family and close friends'.

These reasons for a strong linkage between ethics and family practices might be seen as being quite deeply rooted in history and human experience. This is certainly the way in which they may be presented discursively. But there are other more historically specific reasons why these linkages might seem to be especially strong and this refers to discussions of individualisation and privatisation in modern life. To simplify a complex and contested set of arguments we might say that, with a split between the public and the private, family and intimate life comes to be seen as the source of all that is authentic and all that is bound up with what it means to be an individual subject. In personal, relational and human terms, the focus is increasingly upon family and intimate ties. Home, as the popular phrase has it, is where the heart is.

Thus there are complex, but persuasive, reasons for the linkage between family and ethical practices. Much of this remains at a speculative level but it can be argued that there is no single reason for the success with which these linkages are accomplished. These include the importance of early experiences within family constellations, the links between family and life events and the stress on individualisation and privatisation in modern life. These, and possibly other, influences combine in a complex and overdetermined manner.

To turn to the second concluding question, we need to ask whether the practices approach has any particular contribution to make to debates around families and ethics. Here I wish to draw attention to the emphasis upon everyday or practical ethics that runs through this chapter and the studies that have informed it. 'Morality is not based on sets of principles, but is an activity' (Silva, 2004: p. 60). Silva is referring to everyday ideas of what is 'proper' for parents and children in relation to, in particular, household roles and divisions of labour. These refer to everyday activities, family practices, but are also linked to ethical concerns. Certainly, when individuals are called upon to account for these everyday routinised activities they will respond with some reference to ideas of what is seen as right and proper. This may be highly localised, this is the way *we* do things, and there may be recognition that other people in other families may order matters differently. Nevertheless the sense of the everyday which is a key strand in the practices approach is also at the heart of everyday ethics.

We may round this off with an example. A decision to take on some of the responsibility for the care of an elderly or sick relative who nevertheless wishes to remain in her own home (see many of the examples in Finch & Mason, 1993) may be presented in everyday ethical terms as being based upon a sense of 'the right thing to do'. However, the implications of this decision will include numerous everyday tasks (shopping, visiting, collecting medications etc.) which become incorporated into everyday routines. They are at the core of what we understand by family practices; everyday, routinised and orientated to another family member. Yet this practical ethical understanding of these activities and the decisions that gave rise to them is what provides the linkage between these disparate activities and gives them localised meaning.

There does, therefore, seem to be a close affinity between the practices approach as outlined here and elsewhere and the growing interest in everyday, practical ethics. In both cases we are concerned with links between the part and a constructed whole, between discourses and practices, and the construction and negotiation of meanings. In both cases we are concerned with active processes, with doing family and doing ethics.

There is one final observation that we may make between these affinities between the practices approach and discussions of practical ethics. As I indicated at the beginning, the family and the family practices continue to be the focus of a variety of moralising discourses. In extreme versions, the model is of individuals acting solely in terms of their own interests and gratifications. Any serious consideration of the research dealing with family and family practices should question these accounts. Family practices are routinely linked with everyday ethical decisions and deliberations and individuals are constantly presenting themselves as morally serious individuals.

9
Work/Family Articulation

Introduction

This chapter is slightly different from the ones that have preceded it. While the other chapters have dealt, in different ways, with themes directly associated with family practices and different dimensions of these practices, this chapter may be seen as more of an application of these ideas. Issues of work/family articulation (or, to give the more common term, 'work/life balance') have received considerable attention in recent years in a wide range of countries. The greatest focus has been on mothers who are employed outside home although this concern has extended to fathers and to obligations in relation to family members other than small children.

It can be seen that many of the ideas discussed in the previous chapters have a bearing on this public debate. Thematically, these would include, centrally, questions of time and space but also questions of embodiment, emotions and ethics. Conceptually, several of the ideas discussed in Chapter 3 might be of relevance but the most immediate ones are probably those dealing with 'caringscapes' and 'the total social organisation of labour'.

From the family practices perspective, the issue of work/family articulation is important in that it provides a clear illustration of (a) how family practices extend beyond the confines of the home and how (b) what might be described as *family* practices might also be described in some other terms. Here, the main points of overlap are obviously with gender practices and working practices.

I have noted that the term 'work/life balance' has been most commonly used in connection with the range of issues discussed in this chapter. However, several writers (e.g. Crompton, 2006; Gambles,

Lewis & Rapoport, 2006; Gregory & Milner, 2009; Hochschild, 2003; Ungerson & Yeandle, 2005) have been dissatisfied with this term for a variety of reasons:

- The term 'balance' is unsatisfactory in a variety of ways. It seems to refer to an outcome rather than a continuous process or, as some might say, a struggle. To this extent it seems over-optimistic and the more neutral sounding term 'articulation' (used in Crompton, 2006) might, as is the case in this chapter, be preferred.
- The opposition between 'work' and 'life' sounds a little odd implying, firstly, that the former has nothing to do with the latter. While this may be true for some of the more extreme examples of alienated labour (if they exist) this does not seem to be more generally true. Work is as much 'life' as anything else.
- In practice, the discussion focuses upon 'family' rather than 'life' more generally, and in particular, combining parenthood with paid employment. While there are some other issues to do with work and leisure and some more recent ones to do with work and community or political involvement (both of which deserve more extended treatment) it is probably more advisable to acknowledge that the core concerns are to do with family and intimate life. A broader picture, however, would need to consider community, friends and leisure (Gambles, Lewis & Rapoport, 2006).
- It is also worth reminding ourselves that the term 'work' in this case refers to 'paid employment' (in or outside the home) and that the latter does not exhaust the range of activities subsumed under the former (Pahl, 1984). One only has to consider the extensive debates around housework and childcare as ' work' as well as the less detailed discussion of the work of children in and around the home.

All in all, Hochschild's conclusion seems to be apt: 'work-family balance ... seems to many to be a bland slogan with little bearing on real life' (Hochschild, 2003: p. 198). In this chapter I shall, loosely following Crompton (2006) refer to work/family articulation bearing in mind the difficulties noted with even this modification.

I shall first consider why the articulation between work and family should be considered to be problematic and why there has been a particular concern in more recent years. I shall then look at some of the key issues that have been considered under this broad heading. Finally, I shall discuss how the family practices approach might contribute to these public and scholarly debates.

Why is this a problem?

Discussions of why the articulation of work and family might be seen as being problematic might fall under two headings. The first involves taking a longer-term view of how the key elements of this problem emerged over a period of two hundred years or more. The second is one of considering more recent history and asking why the term 'work/life balance' (or equivalents) has been increasingly used in this period and what particular economic, social or political developments have given rise to this focus.

One of the popular stories about the development of family and intimate life over the last two hundred years or so has been the separation of home from work and the way in which this has mapped upon divisions between the male breadwinner and the female housewife. To describe this as a 'story' is not necessarily to describe it as false (Jamieson, 1998). It has, however, to be recognised that it has been widespread and influential, in a range of public discourses but that it is necessarily something of a simplification that smoothes over numerous complexities. It also has a recognisable before and after narrative structure.

The idea that there might be some problems in the articulation of family and work might have seemed strange in pre-industrial Britain just as it might seem strange in many parts of the world today. Working arrangements, even where the products were sold on the market, were frequently extensions of familial and communal relations. While there might have been some gendered and generational divisions of labour, most family members (plus others unrelated) were involved in the productive process. The use of time and space responded to whatever demands emerged whether these be in terms of the seasons, forthcoming markets, or life-course events to do with marriage and pregnancy.

The development of industrial capitalism and a more urbanised culture entailed the necessary separation of the spheres of home and work. The workplace was separated from the home in terms of time, space and ethos; the disciplines of working life were distinct from those that obtained in the family. Further, and increasingly, this separation of spheres became gendered with the chief earner, the one who travelled between the home and the workplace, being the husband and father. With the development of the male breadwinner we also get the emergence of the housewife and of distinct domestic labour. These distinctions come to apply at most levels of society including the growing professional and industrial middle classes (Davidoff & Hall, 1987). Over time, this distinction came to be reflected in and supported by a wide range of discourses and practices around domesticity, often infused with a construction of the natural order of things.

As I have said this story is not necessarily false although it did smooth over some complexities. These include the number of households where women (and children) continued to be important or main earners, the persistence of family-based concerns and the use of the home itself for some money raising activities such as taking in lodgers or various forms of home work. However, the overall picture was very persuasive, influencing popular representations and scholarly discourses alike. Had this story persisted, in practice as well as in discourse, it is unlikely that issues of work/family articulation would have emerged. The articulation in this story is straightforwardly achieved by the male breadwinner, the one who moves regularly between the two spheres.

The main factor bringing to the fore issues of work/family articulation has been the increasing employment of women, especially mothers, outside the home. Since the Second World War this grew in practice and came to be increasingly supported in public attitudes. In more recent years, the growing proportion of lone mothers within Britain has heightened issues of work/family articulation for this particular section of the population. We might add to this, for growing sections of the population, increasing consumption patterns based upon the home and its maintenance and increasing expectations of the quality of life. Issues of divisions of labour within the home became more complex as care still had to be organised for the children and for other family members.

We can see, therefore, that changes in family organisation and practices have contributed to public pressures for consideration to be given to issues of work/family articulation. There have been demands for state regulation and for changes in the practices of employers. However, this is perhaps to overstate, from a labour–market perspective, the supply side and the demands coming from an increasingly feminised labour force. Other, some might argue more important, pressures, have come from the employers themselves and their demands for greater flexibility in the labour force (Houston, 2005). These pressures and demands are becoming increasingly globalised (Gambles, Lewis & Rapoport, 2006). I shall return to some of these topics in the next section.

We can see, therefore, that the question of work/family articulation has both deep historical roots as well as some more recent influences. Moreover, the concerns are not confined to the UK and the US but have global significance. These influences have contributed to a considerable body of scholarly writing as well as public debate. In the next section

I shall outline some of the key themes to emerge from this literature and these debates before assessing the relevance of the family practices approach.

Articulation of work and family: Key themes

Impact of employment

In much of the literature and debate around this topic it would appear that the main direction of influence is from work and employment to family practices, possibly mediated by gender. There are numerous examples of this influence, which is usually presented as the workplace placing constraints upon the choices available to individual women and, less frequently, men. For example, some women managers might feel that the demands of work make it impossible for them to contemplate having children (Crompton, 2006: p. 72). Certainly it would seem that popular discussions of the ticking away of the 'biological clock' refer to the impact of the perceived demands of a career on delayed motherhood. But the impact, and the direction of that impact from work to home, is seen also in more everyday routines to do with the daily rhythms of life.

It might be argued that the implied model here is one more characteristic of earlier decades where there was a clear structuring of the working day and the working week. This ignores the increasing stress on flexibility where, it might be assumed, the working mother may be able to adjust her working hours to fit in with, say, school hours. While there are examples of this kind of flexibility, it is noted, firstly, that the term itself refers to a wide range of practices and that not all of these have an equally favourable impact on mothers' options.

Booth and Frank (2005: p. 15) list four types of flexibility and other commentators provide similar overlapping lists:

1. The first refers to flexibility in terms of temporary work. This may enable mothers to combine work and parenthood but clearly the timing and availability of this work is dependent upon the employers. Moreover it tends to contribute little to the idea of a career.
2. There is also flexibility in terms of the place of work so that, for example, some of the paid work may be carried out in the home. This may be a solution for a range of activities but also poses new problems for the demarcation of work and family within the domestic space.
3. There is flexibility in terms of working times. In theory this may provide considerable opportunity for achieving some kind of reasonable

articulation but much depends on the actual possibilities that are presented.

4. Finally there is flexibility in terms of the overall number of hours worked. Again there is the possibility of some favourable contract being drawn up but this cannot be guaranteed.

What really matters is the extent to which the flexibility is drawn up primarily in response to the stated needs of the labour force or whether it reflects other economic or competitive pressures on the actual employer. Thus the prospect of flexibility might include the requirement or strong expectation to work 'unsocial hours' or to put in extra hours when the situation demands it. The idea and the reality of flexibility are subjected to competing demands and understandings (Houston, 2005). These competing demands and pressures may be global in their origin and impact (Gambles, Lewis & Rapoport, 2006).

Further the ideals of flexibility may be competing with or subsumed under other pressures within the changing working environment. These are the newer management practices stressing high performance and high degrees of commitment on the part of employees. The implication, often stated, is that career advancement is dependent upon evidence of going the extra mile and being willing to put in the hours (Gamble, Lewis and Rapoport, 2006: p. 50). Clearly this kind of working culture will tilt the balance very clearly against the home and family practices.

It should be noted, finally, that much of the discussion, as Roberts argues, rests upon a quantitative understanding of the idea of time (Roberts, 2008). Hence, many of the strategies adopted in order to achieve some kind of 'balance' revolve around negotiating a reduction of hours or in taking on some kind of part-time employment. Roberts attempts to redress this emphasis by counterposing the idea of 'customisation' in contrast to 'flexibility'. Whereas, as we have argued, the latter term relates, most frequently, to the requirements and practices of employers, the former looks at the individual strategies adopted by an employee. While, presumably, some degree of quantification of time is involved in the process of customisation, the overall emphasis is upon the meaning of time to the individual, the more qualitative aspects.

The role of family practices

If several studies and discussions of work/family articulation seem to stress or to indicate that the direction of influence comes from employment to the home it is likely that this emphasis accords with the way in which it seems to many family members. The demands and pressures of employment might seem, in many cases, to limit the range of options available to, say, working mothers or to fathers wishing to play more active parenting roles.

Nevertheless it would be wrong to see the workplace as the sole determinant of how work and family are articulated. Family practices and the orientations of family members have their part to play. One influential, and controversial exploration of these issues is provided by Hakim in her discussion of women's preferences (Crompton, 2006: pp. 169–70; Hakim, 2000, 2005). Very simply, Hakim argues that women can be divided into three 'preference groupings': 'home-centred', 'work-centred' and 'adaptives', the third constituting the largest group. The first two groupings are self-explanatory; the third consists of women who change their preferences over time in the course of the family life cycle. Similar distinctions can be applied to men but with different proportions.

While these distinctions are described in terms of preferences, we may also see how actual family practices are implicated. In the case of the 'home-centred' women, family practices and the frequently everyday routines of family life and responsibility are seen as primary and any employment has to fit in with these activities and routines established over some period of time. In the case of the 'work-centred', the women may be unmarried or childless and the set of practices that these statuses imply allow for the pursuit of careers or commitments to employment and working cultures. In the case of the 'adaptives' we can imagine a series of on-going, sometimes short-term decisions influencing where the main focus of attention will be. Whatever the decision, we are talking about family practices in the fullest sense since that other family members are taken into account when these decisions are being taken and that the decisions impact upon some of the most routine and everyday activities.

The difficulty with this kind of analysis arises when the basis of these preferences is examined. Are we talking about personality types, about rational calculations of the plusses and minuses or about particular local or family cultures and traditions? When it comes to examining particular cases it becomes increasingly difficult to make clear causal connections. Crompton, for example, points to the somewhat paradoxical situation facing women in France where they are more likely (as compared with some other European countries) to be in full-time employment and where there is a strong tradition of state support for working mothers. Nevertheless the women are not entirely happy with these arrangements, and Crompton suggests that we should look at the domestic division of labour where she finds evidence of 'domestic traditionalism' (Crompton, 2006: pp. 150–62). Put simply, much of the burden of articulating employment and family practices rests with women rather than men. Everyday family practices, the accumulation over time and possibly over generations, of numerous decisions and segregated practices become enshrined in a domestic habitus.

Alternatively, changes (or attempted changes) in family practices may have an influence on this articulation. In a relatively open context in terms of gender relations, men becoming more involved in parenting practices may challenge everyday expectations, may modify their own practices and may seek for changes in working practices (Gambles, Lewis & Rapoport, 2006). These changes, or attempted changes, may affect inter-generational relationships within a particular family constellation. In this case it probably does not say very much to say that a particular father has shifted from being 'work-centred' to being 'adaptive' or 'family centred'; this is just another way of describing the changes. But we need to explore in more detail the complex shift in balances between employment, family practices and state policies.

One thing that these examples demonstrate is that although we are dealing chiefly with parents and their articulations between employment and the home, the significant family others are not limited to the children. Complex negotiations are carried out with the other parent and with other members within the family constellation. These familiar patterns are illustrated in the following quotation:

> The study concludes that a high proportion of call centre employees with children were dependent upon grandparents as primary sources of childcare and/or relied upon partners to 'fill in the gaps'.
> (Hyman, Scholarios & Balday, 2005: p. 123)

Further, the family obligations which are taken into account when articulating employment and family are not themselves confined to the care of children. Consider some of the cases explored by Gambles, Lewis and Rapoport in their multi-country study:

- Zhilah, from South Africa, whose daughter is dying of aids and who is caring for her grandchildren;
- Elizabeth, in the US, whose mother is leaving hospital and going into a nursing home;
- Ravi, in India, with a sick mother. (Gambles, Lewis & Rapoport, 2006: p. xx)

These are just three of the examples explored in this book and these and other examples could be provided from other studies and from everyday experience. In all these cases, the family situation is the prime reality. While work is important, indeed essential to support these routine practices, it has to fit in with these other priorities. The word 'preference' does not really do justice to these experiences.

There is another way in which family practices should be seen as being of fundamental importance. This is in terms of maintaining, policing if you like, the boundaries between employment and family. This is particularly important where the paid work is actually being carried out within the home environment but applies to almost all situations once it is recognised that however much work and family might be separated in theory, they flow into each other in everyday experience. The one may spillover into the other, may constitute some kind of compensation for the other or some degree of clear segregation may be achieved (Dex, 2004: p. 441). An important kind of family practice consists of preventing, as far as is possible, work demands from impinging on the family and vice versa (Cunningham-Burley, Backett-Milburn & Kemmer, 2005: p. 28). This kind of boundary work may be carried on within the home, with other family members, or at the workplace in relation to workmates or colleagues.

In trying to assess the overall balance between employment and family, the list of three inter-related factors that form the basis of Gambles, Lewis and Rapoport's study seems to be especially useful:

1. Work (i.e. paid employment) has become increasingly invasive and demanding;
2. Devoting time and energy to CARE for others and for oneself continues to be of importance 3. The experiences and negotiations between men and women (within the home, within work and more generally). (Gambles, Lewis & Rapoport, 2006: p. 4)

These elements point to themes that have emerged within a wide range of countries; the changing character of work on a global scale, the importance of and the reconceptualisation of care and gender relations. Moreover these themes are interdependent; approach any one of them and you lead to the others.

Gender and class

As should already be clear, discussions about the articulation of work and family were centrally about gender from the outset (e.g. Houston, 2005). More specifically, as is often the case, they were about women. The responsibility for juggling work and family obligations continued to be identified with women and it is women who are chiefly responsible for the 'boundary work' – maintaining some differentiation between the two spheres, within the home. Guillaume and Pochic identify two organisational norms which affect women's (and less frequently, men's) strategies. The first is the 'geographical mobility norm', the requirement to be prepared to move on a temporary or permanent basis. The second is 'the extensive availability norm' which makes claims on an employee's time at or away from the formal place of employment (Guillaume & Pochic, 2009). These norms might seem to be becoming more prevalent with the rise of global companies. Within the overall discussion of how work demands impinge more directly upon women special attention is frequently paid to the difficulties faced by lone mothers. However, as we have seen, family responsibilities do not rest with the care of small children but extend to the whole range of obligations within the wider family constellation.

The discussions, however, have not been confined to women and increasingly the issues faced by fathers are entering into scholarly and policy agendas (Charles & James, 2005). We see this in an increasing range of studies of fathering and fatherhood (e.g. Dermott, 2008; Hobson (ed), 2002) and policy initiatives in terms of more 'family friendly' policies on the part of employers and the extension of paternal and parental leave. Particular attention has been paid to the experiences and practices within Nordic Countries (Brandth & Kvande, 2002; Lammi-Taskula, 2007). One particular focus of interest has been issues of men and masculinity within the workplace. Do men feel inhibited from requesting or taking parental leave through a fear that they may seem out of step with their workmates, that they may seem unmanly or that this may threaten future promotion prospects? These issues provide a good example of the argument that family practices need not necessarily be confined to the home. Nevertheless there is a need for more detailed discussions of men and masculinity in relation to work-family articulation, one not necessarily confined to fathers (Halrynjo, 2009).

We also need to incorporate more recent understandings of gender into this analysis of work-family articulation. This is a more fluid understanding of gender as a process, a question of 'doing' rather than 'being'. In short the emphasis is upon gender practices. Within this kind of approach, gender is not something that exists prior to particular strategies of work-family articulation, something that is, at it were, brought to the work place. Gender identities can also be understood as in part emerging out of these strategies and constructed in the process of articulating work and family (Emslie & Hunt, 2009).

While the focus has been largely on gender, issues of class are increasingly being seen as important. This represents the straightforward fact that the 'choices' facing working mothers and fathers are not equally distributed (Crompton, 2006). Some middle-class or professional employees may have greater bargaining power at the workplace. Moreover, they may be in a better financial position to organise paid child-care. One comparative study of two working contexts comes to the conclusion that 'our data suggest that class mutes the gendered experiences of assistant professors while it exacerbates those of low-income women' (Weigt & Soloman, 2008: p. 622). An example of where class and gender (here seen in global terms) interact is provided by the chains of care whereby middle-class mothers are able to achieve some degree of balance between work and family obligations through the employment of a nanny who herself may well be a mother (Hochschild, 2003; Gambles, Lewis & Rapoport, 2006).

Different models

Crompton (2006: p. 94) notes that while the most frequent household arrangement for couples with children in the UK is the one-and-a-half breadwinner model, this does not apply throughout Europe. Households with two full-time workers are to be found in many other parts of Europe – France for example.

These differences, which are sometimes discussed in terms of different welfare regimes (Esping-Andersen, 1999), reflect different balances between family practices, labour market arrangements and state policies. This triangular relationship between family, market and state has been found to be helpful in examining different patterns of fatherhood (Hobson (ed), 2002) which, as we have seen, is one application of these wider discussions of work-life articulation. We can see that different policies in terms of the provision of parental and paternal leave, the development of 'family-friendly' practices and the provision of nursery or child-care facilities may have important effects on the overall articulation of family and work. But we have also seen that family practices themselves (e.g. the domestic division of labour in French households) may also have a significant role in the final outcome. The elaboration of these more triangular models of family, state and market provides a basis for the understanding of variations and of the inter-dependence of these various influences.

These comparisons serve as a reminder that all kinds of different arrangements are possible even if these may be seen, in some measure, to reflect particular histories and particular cultures. However, comparisons which take the nation state as the basic unit may themselves be subject to limitations. These may be seen internally in, for example, the practices of different ethnic communities. But cross-national differences may also be seen as of increasing importance. These may be in terms of international regulation such as EU legislation. Or it may be in terms of the practices of global, multinational organisations which demand new mobilities and new, flexible, working practices. A Norwegian father describes a sense of 'boundlessness' which may be found in the context of global organisations:

> This weekend for example, when we were on our Sunday walk in the woods, someone phoned, and then I had to open my portable PC and find out something. And with the time difference in Malaysia, which is six, actually seven hours ahead of us ... well it's obvious, isn't it, it's boundless.
> I had a project in Austria earlier on and they work until six or seven o'clock. That's ordinary working hours for them, and I very often had telephone calls when I had come home with the kids.
> (Kvande, 2005: pp. 82–3)

Thus, even in a country which has generous provisions in terms of parental and paternal leave, global working practices and pressures may provide countervailing pressures.

The journey to work

Working time is conventionally understood as the time actually spent at the place of work. Hours of work are usually understood and estimated on this basis. But, as the previous quotations illustrate, this may not take into account the numerous ways in which working demands may impinge upon what is formally family or non-work time. Similar difficulties, of course, may occur when the work, wholly or partially, is carried out at home.

What also needs to be taken into account is the time taken in moving between home and place of employment. These journeys may be undertaken on a daily, weekly or a less frequent basis. Long-distance weekly commuters, for example, may be described as 'new industrial gypsies' (Hogarth & Daniel, 1988). While these mobilities (which might take place between countries) might frequently be accounted for in terms of the availability (or non-availability) of the work for which one was trained, in many cases something more complex is going on.

Something of this complexity is illustrated in this sociologist's personal experience:

> I particularly remember one occasion later on, after my son and I had moved to London, when I scribbled furiously during the whole journey from London.
> On arrival in Leeds the man sitting opposite on the Intercity train commented that I had already done my day's work. And I, while bemused by this everyday deployment of the concept of 'the working day', also recognised that having been up with my son at 5am, and fed and played with him before leaving home, this gentleman's idea of the working day and mine were rather different.
>
> (Wolkowitz, 2009: p. 855)

To some extent the journey to work reflects not only the spatialities of the labour market but also particular sets of family practices. Family practices, as has been argued throughout this book, are not just individual practices but represent courses of action undertaken in relation to family others. Thus for large sections of the population of modern countries, the desire for a home is not simply for a roof over one's head but for a home in the 'right' kind of environment. 'Right' here is shaped in terms of class-based expectations as reflected within and worked out within family practices. In Britain, for example, this might be in terms of being close to the right kind of school and a 'good' environment within which to bring up one's children. In many parts of the world the desire for a 'nice' environment in terms of domestic space, a house with a garden, are important considerations even if the outcome be a longer journey to work than might actually be necessary (Manning, 1978). The important thing to note is that these are family practices.

One may speculate on the overall consequences of numerous family practices and decisions on the organisations of our cities and transport systems. Here we need to consider the part played by the journey to work, a journey reflecting family practices, on the overall articulation of home and work. In some cases the journey may be seen as an extension of work as manifested by the use of mobile phones or laptops on trains. The same phones or laptops, of course, may also be used to keep in touch with family members, as a continuation of family practices. But it is also possible that the journey to work might be seen as a third time/space, one which is neither fully home or fully work although both provide the rationale for being there in the first place and both may impinge at any time. However, this journey might also be the opportunity to develop new skills or interests or to develop acquaintances whose value is that they are just acquaintances and not intimates (Morgan, 2009). At the very least, the journey to and from work, and talk about the journey with all its frustrations and annoyances, might provide a useful buffer between home and work with their different sets of expectations and their different rhythms.

Conclusion

I have said that this chapter represents an attempted application of some of the key ideas associated with family practices. This is not so much an attempt to show some direct connection between the practices approach and various debates surrounding and policies concerning the articulation of employment and family life. Rather, it is an attempt to see how certain practices themes play out in the context of this much-debated topic.

The first point to note is that this is a good example of how family practices can occur some distance, spatially and temporally, from the home. Consider the mother who asks to be excused from a particular meeting at work because last minute difficulties have arisen with her child-care arrangements. Or consider the father who skips the after-hours trip to the pub in order to get home for the children. These and other numerous everyday examples show how family practices, practices carried out with reference to other family members, are enacted away from the home and which involve interactions with non-related colleagues or workmates. I have already noted how the fact that individuals have to travel considerable distances to work themselves reflects another set of family practices relating to decisions as to where to establish a home. Or, yet again, the decision to take a job with a particular employer may be made on the basis of the flexibility of the hours or the closeness to home.

It may also be argued that practices that impact upon individual family constellations may themselves originate in a non-family environment. These may include legislation concerning parental and paternal leave or companies' attempts to develop 'family friendly' policies. Further, companies may deliberately encourage family practices through inviting partners, and sometimes even children, to social events. In some cases, as Guillaume and Pochic argue in relation to the firm they studied, this encouragement may be of 'a certain form of sexual and family life – heterosexuality and marital status' (2009: p. 23).

This whole argument points to the overlaps between the home and the place of employment. While analytical accounts may place these in

separate boxes, the experiential reality is that, in many ways, they constitute a unity. The quotation from Wolkovitz in the previous section illustrates this well; for this particular mother the 'working day' begins with
getting her son up, and feeding and playing with him before working on
the train. All this is before she actually arrives at her formal place of work.
People may adopt strategies, often constrained by personal circumstances
or social structure, in order to separate the two but even while they are
making this separation they are implicitly acknowledging a single reality.

The work of Arlie Hochschild has elaborated some of these ideas
to show how the home increasingly comes to resemble a workplace
(Hochschild, 2003). This may be reflected in the timetables and planning that map out the working week or month or beyond so that key
activities and significant others may be scheduled and co-ordinated.
This may be reflected in the numerous demands that confront a working mother once she re-enters the home on returning from work. In
this case, and contrary to popular representations, the home is far from
being 'a haven in a heartless world'. At the same time, certain aspects
of the employment setting may become more home like, a relatively
supportive adult environment which is a relief from the continuing
tensions associated with real home life.

I have argued that there is value in seeing continuities and overlaps
between work, paid work, and family and that this is the reality for
many people. However, as the previous paragraph indicates, some people want to maintain clear lines of demarcation between the two so
that family concerns do not (insofar as this is possible) impinge upon
work life and problems of work do not spill over into the home. This
apparent contradiction raises two further questions. First, can these two
perspectives be reconciled? And, second, does it matter?

A reconciliation between the two perspectives may be achieved by
drawing upon an old, but still possibly useful notion, of 'central life
interests' (Dubin, 1962). The idea, which is spelt out in some detail in
Dubin's original article (and which has echoes in Hakim's more recent
discussion of preferences cited earlier in this chapter), is more or less self-
explanatory. While these life interests tend to be dichotomised in terms
of family and work there is no reason why they cannot be extended to
include friendship, travel or spiritual enlightenment. Further, while it
is likely that these interests may prove to be relatively stable over a life
course, they do not necessarily rule out the possibility of change in the
life of an individual. Finally, while a dichotomous division between
central life interests in terms of family and work might be mapped on to
divisions between working class and middle class and between women

and men this is never been straightforward and is likely to be even less so in the twenty-first century. What this means is that the overlapping worlds of home and work become even more complex with numerous criss-crossings of different interests, career paths and commitments all of which may be subject to change over time. Those for whom home and family relationships are central life interests, work will be a means to an end and there will be a clear demarcation between home and work. For those who have a clear notion of a career or for whom work and workmates are central, family obligations and commitments will be of lesser importance. Some will wish to maintain clear boundaries around the two spheres, while others may accept or even welcome the continuities in everyday life. The discussion of family practices may, along with other concepts and ideas, help analysts and policymakers find a way through this late modern tangle.

The discussion of the articulation of home and work, then, provides a good illustration of how family practices may occur away from the physical home. We can also see, clearly, how family practices overlap with other practices. In many of the examples that I have given, what I have described as 'family' practices could equally be described as gender practices, doing masculinity or femininity, or working practices. We have also seen how class practices (and doubtless several other practices) are also involved. This overlap between family and other kinds of practices is not, I argue, a weakness of the overall approach but one of its distinctive strengths.

But, to return to the second question, does it really matter? I have argued that the practices approach provides a particular, but not an exclusive, point of entry into these debates. By stressing the importance of family practices and the way in which they play a major role in constructing everyday life we may realise that individual men and women are not simply reacting to the necessity for earning a living and the ordering of the work environment. They may make particular demands upon the working environment in the name of family practices or may develop innovative ways of responding to the labour processes. The 'family' that provides the basis for these practices, whether reactive or innovative, may not simply be a couple and children but may consist of several generations or a wider family configuration. In observing practices around the work/family divide within communities and over generations we may come to understand processes of structuration. Thus, if employers and employees alike assume that there are informal networks of child-care arrangements within a particular community, then they may proceed on this assumption. If, for all kinds of reasons, these networks become weaker, then these assumptions may fall out of step with reality and new patterns may emerge. At its simplest, the practices approach, serves as a reminder that we are rarely, if ever, dealing with individuals and individual decisions.

In presenting this argument I am also aware that these discussions point to possible limitations of the practices approach. Much of what I have said in this chapter might also be said in terms of the Total Social Organisation of Labour (TSOL) approach or 'caringscapes' or the configurational approach. My argument in this book is that these are not competing theoretical orientations but provide a range of overlapping perspectives on some of the key issues of our time. Where the emphasis lies will, ultimately, depend upon the level of analysis and the kind of problems being tackled as well as the preferences of the researcher. Some of these issues, as well as the place of ' family' within the overall frame of analysis, will be discussed in the next and final chapter.

10
Conclusion

Introduction: The aims revisited

In Chapter 1, I described the aims of this book in these terms:

- To explore the range of meanings and usages attached to the term 'family practices';
- To discuss the relationship between the term 'family practices' and the more general use of the word, 'practices';
- To consider possible ambiguities within and criticisms of this term and its usages and to explore how these might be met and developed in order to enhance our understanding of family life and its place within wider social contexts.

In the course of the discussion, and especially in Chapter 3, a fourth aim emerged. This was linked to the third aim but deserves separate consideration and this was to do with the place of the family practices approach alongside some other alternative approaches that were being developed at around the same time.

In terms of the first aim, the exploration of the range of meanings and usages I think that little more needs to be said on this topic as it is one that I have referred to in different ways throughout the discussion. However, it is worth noting a general distinction between the use of the words 'family practices' as a descriptive term and its use as a concept. Several studies have used the term 'family practices', some with reference to my formulation and others without. Whether or not they refer to my original formulation is of little significance since, as I argued, an overall practices approach has been present within sociology for some time and so that the application to family living was probably inevitable. When the words are used as a simple term of reference they do little more than point to a broad area of enquiry, one which is roughly concerned with what people do rather than with the structure of the institutions within which these activities take place.

My original intention, elaborated here, is to go beyond this simple descriptive usage and to try to explore what is implied when the term 'practices' is used. I listed several inter-related themes in order to conduct this exploration. These included a focus on the everyday, a focus on doing or action, a stress on the fluidity of boundaries, and emphasis on reflexivity and so on. These themes did not, taken together, constitute anything like a formal model but did seem to fall together in the process of constructing any overall way of thinking about family relationships, one that seemed especially appropriate for the late twentieth, early twenty-first, centuries.

Subsequent scholars have, in the spirit of the relatively open-endedness, taken different themes from this articulation of what was entailed in the term 'family practices'. Very generally, it would seem that what proved most influential was the emphasis on 'doing' family and that way in which family was continually being constructed and reconstructed through the performance of these practices. In the course of this present re-exploration I have, similarly, drawn upon or emphasised different themes at different points. What matters is the overall orientation to the study of family living rather than any attempt to provide a new theory of family life. Put in another way it is an attempt to theorize, to think theoretically about, family life rather than to construct a theory. To this extent my overall emphasis on the process of theorizing rather than an accomplished theory is consistent with my overall fluid and open-ended approach to family living itself.

The second aim was to explore the relationship between discussions of practices in general and family practices in particular. As I argued, the term 'practices' was already widely used when I embarked upon this investigation and a body of work sometimes known as 'practice theory' is now well established (Reckwitz, 2002; Schatzki, 1996). In Chapter 2, I explored this relationship and established there was some broad congruence between the general and the particular. This conclusion is not so banal as it might seem, since I worked backwards from my usage of the term 'practices' to the more established and more general usages. Had it been the other way, the 'family practices' approach might have been presented as a simple application of these more general, and already current, approaches. When I was elaborating the idea of family practices I was only vaguely aware of these more general orientations.

My development of the idea of family practices was not, therefore, a simple application of this more general body of theory. It also raised questions as to what is distinctive about *family* practices as opposed to these more general usages. I concluded, initially, that family practices were those practices that were orientated to others who were designated as, treated as, co-family members. They might also be orientated to, on certain occasions, constructions of the family, or our family, as a whole as in the case of family display (Finch, 2007). In this way, therefore, family practices are reflective practices; in being enacted they simultaneously construct, reproduce family boundaries, family relationships and possibly more discursive notions of the family in general. Drawing up a will, for example, may be a family practice as it simultaneously has a practical and a symbolic aspect, defining particular relationships and their significance.

This more explicit exploration of the relationships between 'practices in general' and 'family practices' also raises the question as to whether the former may gain from the explorations of the latter. In other words, as I have suggested, my work was not simply an application of some general principles but an exploration and an elaboration of these principles. There are at least a couple of areas where, I argue, practice theory in general might benefit from this more particular investigation.

The one is the articulation of a possible tension between practices as action, as doing, and practices as something more routine or habitual. Of course, simply because something is routine or habitual does not mean that it is not a social action. What makes it social action is that it is orientated to others, their expectations and our expectations of their expectations and that it is accountable. In other words, when questioned, an individual might always be able to account for the action even if the response is simply along the lines of 'that is the way we have always done things'. Hence it is not simply instinctual or reactive.

Take, for example, the celebration of Christmas. Autobiographical accounts are full of descriptions of how and when gifts were given and opened, cards displayed and so on. They are presented as something repetitive, year after year, and unchanging. Memory may, of course, blur over changes and differences over the years but the presentation is often in terms of 'this is the way we celebrated Christmas'. This sense of repetition was often reinforced by a diffuse sense that other people were more or less doing the same kind of things.

However, when individuals leave their parental home and, at some stage, enter into partnerships or marriage, they discover that other people have other ways of celebrating Christmas. These differences might be quite subtle but may have emotional significance. What was more or less taken for granted now becomes a matter for more specific negotiation. This will be even greater where the partners belong to different faiths or cultural traditions. Discussions about 'the way we do Christmas' are, of course, part of a much wider set of negotiations entailed when two different biographies become more closely entwined. This is part of what Berger and Kellner mean when they described marriage as a 'nomos creating' institution (Berger & Kellner, 1964).

What, then, are the differences between 'practices as action' and 'practices as habit'? This is, I argue, a matter for further investigation but the following points would seem to be relevant:

- The former are more explicit and more likely to be the subject of open, rather than covert, negotiation;
- The former are more likely to be both accountable and called to account, that is individuals are more likely to find some need to justify the particular course of action taken;
- The latter, 'practices as habit', are more embedded within relatively close-knit networks of relationships or community ties such that these relationships and ties provide immediate, if unspoken legitimation for whatever action might be taken;
- The former, 'practices as action', are more likely to come to the fore in circumstances of change or novelty.

To talk about a tension between these two ways of conceptualising practices is to overstate the case. It is more likely that we are talking of a continuum rather than a sharp opposition. Further, it is a distinction which is rooted in classical sociological theory (traditional and rational action) as well as one that has its applications in family practices. Askham's distinction between identity and stability in relation to marriage might be said to draw upon a similar contrast (Askham, 1984). It may, however, be argued, that with increasing mobilities of all kinds, 'practices as action', at least in a family context, are likely to grow in relatively significance.

This discussion is one that emerged, in my thinking, in the process of moving between 'practices in general' and 'family practices'. Another more direct way in which a discussion of family practices might illuminate these general concerns relates to the argument that what may be described as *family* practices might equally be described in some other way such as gender practices, generational practices and so on. This understanding emerged out of the realisation of the fluidity of familial boundaries not only in terms of who, at any one point of time, counts as a member of a particular family but also in terms of the ways in which family life necessarily overlaps with other key areas of social life.

There are, necessarily, various modifications and adjustments that are made as we move between practices and general' and 'family practices'. Family practices have their own points of reference and distinct colouring but analysis in these terms can still be seen to derive from, and possibly contribute to, these wider discussions.

The third main aim of this book was to explore any possible criticisms of the family practices approach and to see to what extent any modification of this approach might be required as a result of these criticisms. I identified four main areas of criticism, some of which are more potential than actually realised criticisms:

- The heteronormativity implied by the emphasis on 'family' in family practices. I think there is some justice in this criticism although it is one which might apply whenever the term 'family' is used in social enquiry. Broadly speaking my response is to acknowledge the limitations and to stress that I am dealing about family rather than intimate practices or personal life. Further, family life and practices can be seen as cutting across different groupings defined in terms of their sexuality. I shall return to this later in this chapter.
- The underplaying of structural constraints and historical context. I argue that this might be true in terms of my presentation but I hope to show that this is not necessarily a fatal flaw.
- The playing down of the importance of discourse. Again this was probably true in the original formulation but, again, can be accommodated in a wider discussion.
- The playing down of the darker sides of family life. This, again, may be true in terms of my particular treatment of the topic but it is not necessarily the case. Family practices may be damaging, indeed fatal, and unwelcome as well as the reverse.

On the whole, I think that while there are some critical points that can be made about the family practices approach most of them can be accommodated within the original framework. This is because it is not so much a theory or a model but a general orientation which may provide the stuff from which theories or models might be constructed.

I have stated that a fourth aim emerged in the course of this investigation and that was to consider the family practices approach alongside certain possibly competing approaches. Some of these had several points of overlap with the practices approach: these were studies of intimacy, personal life and family configurations. All these shared a recognition that a concentration on 'the family' (especially models based on the nuclear-family-based household) were increasingly seen as highly limited and as failing to capture complexities of personal and intimate relations in modern life. Family practices were included within these other orientations but these orientations were not limited to family relationships as conventionally understood. I have increasingly come to recognise the importance of these overlapping but distinct approaches but, as I shall argue later, I feel there is still a place for the word 'family' and the study of family practices.

The other 'alternatives' were slightly different: the study of caring-scapes and of TSOL. Here, as with the other alternative approaches, family relationships and practices were included within something wider but these wider fields went beyond family, personal and intimate relationships to explore work and employment, neighbourhoods and the impact of the state. These approaches were important in that they addressed slightly different problems in which family relationships were frequently implicated but it was the way in which these practices were articulated with other areas of social and economic life that were important. Thus key areas of social life such as caring and work required attention to families and households but also focussed upon the wider sets of networks and institutional connections within which domestic life is placed. In all these possible alternatives, however, we are trying to 'place' family practices within something else that overlaps with or includes these relationships. The importance of family studies, as I have argued earlier (Morgan, 1996) is not simply in terms of the particular relationships subsumed under the label 'family' but also in these points of overlap, articulation and inclusion with other relationships and other practices.

Other connections

After the earlier chapters in which I outlined the original argument and its origins, explored possible criticisms and attempted to place my argument alongside other approaches, I moved on to a set of specific topics. These were time and space, the body and embodiment, emotions and morality. To what extent are these relatively arbitrary chapter headings or are there some connections between them?

At a relatively superficial level we can argue that these topics are all relatively recent arrivals within sociological analysis. In textbooks with which I was associated from the 1960s onwards, these words could not be seen in the indexes let alone in the list of chapter headings. In a fairly recent collection called *Developments in Sociology* (Burgess & Murcott, 2001) there are some passing references to the body and morality in the index and my chapter on family studies refers to 'the moral economy' and 'the emotional economy' but none of these topics deserves a chapter to itself. But from the late 1990s onwards, these topics have grown in importance within the discipline as a whole. While these topics appear to be disconnected from each other, each book or edited collection dealing with them will begin in a similar way. The apparent neglect of the topic under discussion will be highlighted although this will be followed by attempts to show that there are some continuities going back into the sociological canon.

To some extent, therefore, I am simply recognising these developments within sociological analysis and attempting to apply them to the study of family practices. I realise, as I write that there are other developments within the discipline that I do not recognise in this way; mobilities (although there are passing references), the physical context, globalization and the environment, for example. Simply to list these omissions is to recognise the constantly changing notion of sociological enquiry. Yet I feel that my choice of these headings is not entirely arbitrary and that I am not simply trying, and failing, to catch up with recent developments in a rapidly changing field. There are some reasons for these choices even if I was only dimly aware of it at the time.

I have attempted to show that the study of family practices can be enriched by taking account of developments in these apparently separate fields and that, possibly, these developments might themselves be enhanced by taking more account of family practices. This is not to say that there is some exclusive claim being made on the part of family studies here. These developments could equally be applied to the study of work and employment (Morgan, Brandth & Kvande, 2005), to social class, to leisure or to numerous other areas of social enquiry, recent or well-established. However, I do argue that there are strong connections between these new developments, taken separately, and the study of family practices.

Without repeating any of the arguments in detail, consider the following:

- In the study of family practices we see the interplay between different understandings, constructions and experiences of time; time as a negotiated resource, cyclical and repeated time, historical time.

- Similarly, we see links and overlaps between practical, symbolic and imaginary space and also between space and place.
- The different ways in which we can speak about the body and embodiment are played out in family practices.
- Emotion work and emotion management are readily and strongly associated with family practices.
- Much family talk and family practices almost inevitably deploy the language of morality and are linked with ethical practices.

I would argue that these are all deep connections. Although considerations of space and time, the body, emotions and morality were frequently absent from many accounts of family life (including my own) once they are introduced, the connections seem obvious and inevitable. Further, as I attempted to show, the application of these areas of enquiry to family studies is not only to the benefit of the latter but also of the former. Thus most, if not all, of these areas of enquiry benefit from being seen through family spectacles if only because they remind us of relations and relationality.

One of the benefits of linking these developments to the study of family practices lies in the fact that in so doing we became aware of connections between and across these different developments. As Gabb's innovative study shows us, there are clear links between domestic space, emotions and embodiment (Gabb, 2008). It would not strain the analysis too much to explore links between these intimate spaces and practices and issues of morality or time. This is not to argue for some huge synthesis, the end product being a diagram with several boxes and lots of inter-connecting arrows. In many cases, the interconnections come naturally. When thinking about embodiment we think about bodies in relationships (which necessarily introduce questions of morality and emotions) within social space and time. When thinking about morality we are quickly brought (as Nussbaum (2001) so clearly demonstrates) to questions of emotions and embodiment. What is important is to develop a general sensitivity to these connections and to explore them in relation to whatever particular topic we happen to be concerned with: work/life balance, domestic violence, environmental issues or whatever.

Methodological issues

Are there any particular methodological issues associated with the practices approach? While, in principle, almost any tools of social enquiry might be deployed in the study of family practices, in reality it would seem that the approach more readily leads in the direction of qualitative analysis. It would seem that the key terms of the practices approach – doing, relationality, fluidity – would almost inevitably demand some form of qualitative study. And, indeed, most of the studies cited in this book that make some use of the idea of family practices are strongly qualitative in emphasis.

But qualitative analysis itself covers a wide range of methods and techniques of social enquiry. Further, some forms of qualitative enquiry may not necessarily provide for an adequate realisation of the practices approach. The qualitative interview (in-depth, open-ended, or whatever) despite being widely used in family analysis might not necessarily tap into family practices. The relationship between what is said in an interview situation and what is actually done (the core of a practices approach) remains a complex issue within qualitative family studies. Part of the resolution of these issues lies in recognition of the fact that interview talk is itself a form of family practice, a way of constructing the family self in a particular context. Clearly a degree of reflexivity on the part of the researcher is required here as in other approaches. The study of doing, the core of the practices approach, might suggest some form of observational techniques but these themselves present a whole host of practical and ethical problems.

An example of the innovative combination of methods is provided by Jacqui Gabb (2008) who also includes a comprehensive overview of the methods available to family researchers. Rather than go through the whole range of methods that might be deployed in exploring family practices, I shall focus here on the potentialities of narrative and auto/biographical research. The 'turn' to such methods has been well-explored in recent texts (Chamberlayne, Bornat & Wengraf, 2000; Roberts, 2002) although the term indicates a variety of different techniques and scholarly traditions. These include:

- Oral histories;
- Life histories;
- Life-course analysis;
- Qualitative interviews;
- Critical reading or re-reading of existing auto/biographical texts, published or unpublished;
- Memory work;
- Collective writing projects;
- Use of a range of visual material and representations including photographs, maps and networks, masks etc.

What all these different techniques have in common is perhaps difficult to discern but it is possible to see a loose cluster of common concerns emerging through the work of a variety of individual projects and activities of study groups:

- A broad concern with individual lives in time as opposed to, say, attitudes collected at a single point of time.
- A concern with linking these individual lives to broader historical trends and movements.
- A recognition of the interdependencies that exist between research and researched. This is reflected in the use of the compound term 'auto/biographical' in some cases (Stanley, 1995).
- Exploration of the distinction between and the relationships between 'the life lived' and the 'life told'. This may be a difficult distinction but it points to the need for constant critical examination of the processes of gathering stories.
- A concern with a whole range of lives and a whole range of experiences. The contrast here is with the lives of celebrities or other well-known figures that are popular in the publishing world.

Of course, not all auto/biographical studies will attend to all of these considerations in equal measure. Nevertheless they provide some broad points of convergence.

The relevance of this broad set of techniques and approaches for the study of family practices should be clear. The first is the obvious temporal awareness that is so central to family life and relationships. For example, if one wishes to ask why it is that a particular family member (frequently a woman) finds herself with particular caring responsibilities, it is important to look at how her family relationships have developed over time (Finch & Mason, 1993). We need to see how certain expectations and moral identities are established through numerous exchanges and occurrences over the life-course, or that section of the life-course which is especially relevant in this case.

The second is a concern with 'doing' family rather than, simply, attitudes about or orientations to the family in general or to a particular family constellation. These attitudes are important but they are not the main focus of concern. We are more concerned with what people do, when they do it and in relation to whom. This can be established through careful qualitative interviews or the use of research diaries or visual material. The apparent triviality of some of these activities is of no concern.

The third is the desire to provide some linkage between the everyday experiences recounted and accounted for and wider patterns of historical and cultural change. Often the simplest pieces of information can begin to suggest these connections. To state, as my passport does, that I was born in London in 1937 is already to suggest a range of historical experiences for further investigation; the Second World War, the establishment of the welfare state, post-war austerity, national service and so on. Making such connections in a persuasive and non-arbitrary way is not as simple as this might suggest, but the task is aided by dealing with substantial time periods.

Fourthly, we may see the constant interplay between researcher and researched in the conduct of auto/biographical enquiry. What is important is not simply what is told but how it is told and how particular stories are elicited from research subjects but researchers who are themselves located in the shared flow of time. How is a particular narrative produced and how is it represented on the page in the book or journal article? What kinds of readers are anticipated and how are potential readers constructed in the course of presenting an auto/biographical account? These are complex questions and their complexity should remind researchers and readers alike that this mode of qualitative research is not an easy option.

Finally, the auto/biographical approach does allow for a certain fluidity and openness. For one thing the practitioners may be found in a variety of disciplines. For another, everyday accounts of lives, properly elicited, do not always make hard and fast distinctions between particular areas of life. An account of how a mother spends the first three hours of a weekday, for example, will move between family, employment, education and several other spheres.

I have simply indicated that the auto/biographical 'turn' may be seen as having particular affinities with the family practices approach. I am not, however, claiming a necessary connection between an overall approach and particular research techniques. For one thing, as we have seen, the 'auto/biographical turn' denotes a variety of different and not always immediately compatible techniques. Some forms of life-course analysis, for example, can have a strong quantitative orientation. Further, other more established approaches may well yield important insights into family practices. For example in Oinonen's study, *Families in Converging Europe* (2008) there is a chapter called ' On Family Practices'. In this case, 'practices' are contrasted with 'ideologies' and both are dealt with through extensive use of quantitative data. In the practices chapter we have demographic data (marriages and remarriages, cohabitation, divorce, births) and statistics relating to the employment of women. While most studies using a practices approach would probably use more qualitative material looking at, say, interactions between mothers and children rather than births, it is nevertheless important to remember that these readily available statistics represent family practices and can be, perhaps should be, the basis on which further analysis can proceed.

What is so special about *Family* practices?

An issue that I have returned to at several points in this book is about the distinctiveness, or otherwise, of *family* practices as opposed to both practices in general and other sets of practices which might seem to be more inclusive or more relevant to a changing society. Thus I have considered the possible claims on behalf of intimacy and intimate practices, personal life and configurations, family or otherwise. There might seem to be good reasons, theoretical and political, to drop the term 'family'.

One reason why there is still concern about the theoretical and political connotations attached to the use of the term 'family' may lie in a familiar confusion between 'family' and 'household'. The kind of model which gives concern is a particular combination of the two, a household which consists of the standard model family consisting of heterosexual parents and young children. It is no difficult task to show that, statistically, such a model represents a minority of all households. However, once it is remembered that the adult members of this household themselves have or have had parents and may well have siblings and other family relationships extending, possibly, over the world. Thus, shifting attention away from a particular kind of household to the wider family network or constellation introduces us to arrangements that can include almost anything that might be seen to be characteristic of 'late modern' intimate relationships; lone parents, re-constituted families, gay partnerships, single-person households and so on. The term 'family' here refers not to any particular units but to the links between these units. Some of these links might be dormant or unrecognised and certainly not all of them will be active all the time. Nevertheless they represent potential ties and even where they are ignored or neglected it is frequently the case that this neglect has its origins in the history of a particular family network.

Against this it might be questioned why I focus on relationships which are defined as 'family' (however far-flung in some cases) rather than on other relationships, such as friendships, which might be of equal or greater importance to particular individuals? Certainly it is possible to devise networks or overlapping circles which include friends and acquaintances as well as family ties (Spencer & Pahl, 2006). Certainly, also, there is frequently some overlap between ideas of friendship and family. When asked to name a 'best friend' individuals may, for example, cite a parent or a partner. At the same time there is also evidence to suggest that individuals do distinguish between family and friends (while recognising the importance of the latter) and have slightly different expectations from members of each grouping. How people distinguish (if they do) between friends and family and their relative significance to the individuals concerned are matters for empirical investigation. The argument that friends are chosen while family is, for example, given is a fairly common elaboration.

Therefore, a preliminary answer to the question 'what is special about family practices?' is to recognise that for some people under some circumstances they are not all that special. This may reflect choice or circumstances such as major social or personal disruptions. Nevertheless this response also implies that family practices are, indeed, special for, at least, some people.

This was one of the points of departure for the development of the idea of family practices. While accepting all the criticisms of family life and, in particular, of constructions which appear to present a normative version of 'the family' it is also the case that family life continues to be important for many people for much of the time. Further, even at the level of discourse and representation, matters are not so straightforward. While there continue to be many political or popular representations of a particular version of family life based upon heterosexual marriage there are, equally, popular accounts which present a more complex and finely nuanced perspective in which multiple partnerships, bi-nuclear households and gay relationships coexist with more 'conventional relationships' (a lot of soap operas and sitcoms for example). In discursive or ideological terms, competing versions of what family and intimate relationships are or should be about are to be found.

I am more concerned here with the importance of family in everyday life, the way in which family matters enter into everyday conversations. Here, I should argue, the emphasis is nearly always highly focussed on particular relationships at particular points of time. It is rarely on the idea of family in general. I am talking about conversations about a parent's terminal illness, about a daughter's achievements at university, about the break-up of a brother's marriage. In the course of these conversations individuals may show despair, pride, pleasure or frustration but, whatever the emotions expressed, the underlying implication is that these are important or serious matters, stories that deserve the telling.

Why do we find this importance? Part of the answer must be that family life is so readily and so easily linked to the everyday (Morgan, 2004). There is a taken-for-granted, given quality of family relationships which seems to rule out choice in the matter. Single adults and childless couples may still feel excluded from a particular club based around the themes of coupledom and parenthood. And even if one perceives one's present intimate status as one that is freely chosen, this does not eliminate the fact that one has or had parents and siblings whose presence is not easily dismissed.

This given or inevitable quality of much family life in part accounts for the centrality of family practices. But partly as a consequence of this, there is also the way in which, as I have argued at different points, family life so often seems to bring together different strands which seem to be closer to what life is really about. To this extent there is an over-determined quality of family life. Any one influence might account for its significance but we are dealing with several influences and their various combinations. Here I return to the chapter headings in the main body of this book, referring to time and space, the body and embodiment, emotions and ethics. I have argued that this is not entirely an arbitrary collection of fashionable topics but these which, in their different ways, have strong connections with family practices and, through these, with each other.

In order to illustrate this let us take one of the possible topics of conversation mentioned earlier, a parent in the last stages of a terminal illness. Very schematically, I can show how these different themes are implicated:

- Space. Where is the individual concerned to be located in the last months or weeks of her life; at home (whose home?), in hospital or in a hospice? How is care and concern to be expressed and mobilised over a set of widely dispersed family members and others?
- Time. How does the expected timetable associated with dying impact on other timetables? What impact does the impending ending of a life have upon the wider family constellation? How are different memories brought to the fore at this particular time?
- The body and embodiment. This is clearly a dominant concern here in ways which were discussed in this particular chapter.

- Emotions. Again, this scarcely requires emphasis. Such events are the occasion for the expression and management of a complex set of emotions, one's own and others, within a family constellation.
- Ethics. Again such events raise questions of the 'right thing to do' and on the obligations assumed by or placed upon a particular family member.

All these influences are important both singly and in their various combinations. It might be possible to use these themes as elements within a larger systemic model but that is not my aim here. It is to underline my argument that the importance of family life is not simply the product of bourgeois or patriarchal ideology (although these play their part) but arises in very complex ways out of everyday life itself. Not all of these elements will be equally important in each circumstance but the sheer weight of these influences taken together must be recognised.

Let me, in conclusion to this section, make it clear what is not being said. First, I am not arguing that other sets of intimate relationships (especially friendships) are to be downgraded. Their importance in particular intimate constellations or personal communities must be determined on a case-by-case basis. Second, I am not arguing that family life is always desirable or to be welcomed. The same or similar combination of elements could be produced in, for example, a discussion of child abuse or neglect. And, even in the case selected, individuals within the family constellation may experience considerable degrees of ambivalence as the mother's life comes to an end. Finally, in attempting to answer the question as to the overall importance of family practices (as opposed to other practices) I am not seeking to establish a moral hegemony of family life. I am seeking to explain, recognising the historical and cultural limitations of any explanations that may be provided. In explaining and understanding it is to be hoped that readers will gain an understanding of what can be changed and where change is very difficult.

Conclusion: Where next?

I hope that I have shown, in the previous chapters, that the family practices approach still provides a good way into thinking about family life, if not family life in general, then at least family life in late modernity. I have shown that other scholars have found the ideas useful and I have attempted to show how they can illuminate particular concerns with themes such as embodiment, ethics, emotions and so on. I hope that scholars will continue to find these ideas useful and will seek to apply and develop them in their particular enquiries.

In the course of this discussion in Chapter 3 I identified one approach which represents a possibly fruitful development of the practices approach. That is the idea of 'family display' (Finch, 2007). This represents an explicit development of the practices approach and it is one that has attracted considerable attention and is already beginning to inform family inquiry.

I have also noted some other approaches that developed around about the same time and which sometimes include references to family practices. These include the study of intimate life, of personal life, of family configurations and, somewhat differently, caringscapes and the total social organisation of labour. I have argued that these are not necessarily alternatives but can inform each other in all kinds of ways. The idea of family configurations, for example, points to patterns of inter-connections within family networks and the ways in which these handle external pressures and constraints while, themselves, being a source of new constraints. Further, while these configurations may begin with and encompass family relationships they may also extend outwards to include significant, but non-family, others. Family practices here refers to what goes on within these configurations, the everyday and not-so-everyday activities that define who is within these configurations and mark out boundaries and distinctions within them.

In the case of intimacies and personal life I have suggested that these are not simply more general terms which include or encompass family relationships but rather overlap with family practices. Thus, some family practices may not be especially intimate in any sense of the word, and 'personal communities' may include some non-family members while excluding some family members. While there are many points of overlap and connections between the study of intimacy, of personal life and family, these areas can be studied separately while also being aware of points of overlap and articulation. Furthermore the practices approach can be extended, with little modification, to the studies of intimacy and personal life.

In the cases of caringscapes and TSOL, we find different areas of overlap and distinctness. Clearly, these two sets of evolving ideas are more general than family practices (the mobilisation and the allocation of care and the articulation between all kinds of working practices and divisions of labour) and raise all kinds of different issues. But their elaboration serves as a reminder both that family practices cannot be easily removed from other sets of practices and that these wider issues of care, work, class and gender can never be fully understood without looking at family practices.

Within these various and other approaches I feel that at least one core message to be derived from the practices approach will find a place; the idea that family is something that people 'do' and in doing create and recreate the idea of family. Sometimes this sense of family will be distinct from and possibly in opposition to other structures and practices. At other times there will be a relatively easy sense of flow between these and other areas of social life. So long as some people continue to think that, from time to time, family life is in some ways distinct, there will be a need for the practices approach.

References

Adam, B. (1995). *Timewatch: The Social Analysis of Time.* Cambridge: Polity Press.

Adams, J. (1994). The Familial State: Elite Family Practices and State-Making in the Early Modern Netherlands. *Theory and Society*, 23(4) 505–40.

Ahlberg, J., Roman, C. and Duncan, S. (2008). Actualizing the 'Democratic Family'? Swedish Policy Rhetoric Versus Family Practices. *Social Politics*, 15(1) 79–101.

Askham, J. (1984). *Identity and Stability in Marriage.* Cambridge: Cambridge University Press.

Backett-Milburn, K. (2000). Children, Parents and the Construction of the 'Healthy Body' in Middle-Class Families. In A. Prout (ed.) *The Body, Childhood and Society*, Basingstoke: Palgrave Macmillan, 79–100.

Banfield, E. C. (1958). *The Moral Basis of a Backward Society.* New York: The Free Press.

Becher, H. (2008). *Family Practices in South Asian Muslim Families: Parenting in a Multi-Faith Britain.* Basingstoke: Palgrave Macmillan.

Beck, U. and Beck-Gernsheim, E. (1995). *The Normal Chaos of Love.* Cambridge: Polity.

Beck, U. and Beck-Gernsheim, E. (2002). *Individualization.* London: Sage.

Berger, P. and Kellner, H. (1964). Marriage and the Construction of Reality. *Diogenes*, 46, 1–23.

Bhambra, G. K. (2007). *Rethinking Modernity: Postcolonialism and the Sociological Imagination.* Basingstoke: Palgrave Macmillan.

Bonvalet, C. and Lelièvre, E. (2008). Entourage: A Web of Relationships in Reference Spaces. In E. D. Widmer and R. Jallinoja (eds) *Beyond the Nuclear Family: Families in a Configurational Perspective*, Bern: Peter Lang, 375–96.

Booth, A. L. and Frank, J. (2005). Gender and Work-Life Flexibility in the Labour Market. In D. M. Houston (eds), *Work-Life Balance in the Twenty-First Century.* Basingstoke: Palgrave Macmillan, 11–28.

Bourdieu, P. (1977). *Outline of a Theory of Practice.* Cambridge: Cambridge University Press.

Bourdieu, P. (1990). *The Logic of Practice.* Cambridge: Polity.

Bourdieu, P, (2008). *The Bachelors' Ball: The Crisis of Peasant Society in Béarn.* Cambridge: Polity.

Brandes, S. (2001). The Cremated Catholic: The Ends of a Deceased Guatemalan. *Body and Society*, 7(2–3) 111–120.

Brandth, B. and Kvande, E. (2001). Flexible Work and Flexible Fathers. *Work, Employment and Society*, 15(2) 251–67.

Brandth, B. and Kvande, E. (2002). Reflexive Fathers: Negotiating Parental Leave and Working Life. *Gender, Work and Organization*, 9(2) 186–203.

Bryceson, D. and Vuorela, U. (2002). *The Transnational Family: New European Frontiers of Global Networks.* Oxford: Berg.

Burgess, R. C. and Murcott, A. (2001). *Developments in Sociology.* Harlow: Pearson Education.

Carmichael, K. (1991). *Ceremony of Innocence: Tears, Power and Protest*. Basingstoke: Palgrave Macmillan.

Chambers, P., Allan, G., Phillipson, C. and Ray, M. (2009). *Family Practices in Later Life*. Bristol: The Policy Press.

Chamberlayne, P., Bornat, J. and Wengraf, T. (2000). *The Turn to Biographical Methods in Social Science: Comparative Issues and Examples*. London: Routledge.

Chandler, J. (1991). *Women Without Husbands: An Exploration of the Margins of Marriage*. Basingstoke: Palgrave Macmillan.

Charles, N., Davies, C. A. and Harris, C. (2008). *Families in Transition: Social Change, Family Formation and Kin Relationships*. Bristol: The Policy Press.

Charles, N. and James, E. (2005). Gender, Job Insecurity and the Work-Life Balance. In D. M. Houston (ed.) *Work-Life Balance in the Twenty-First Century*. Basingstoke: Palgrave Macmillan 170–88.

Cheal, D. (2002). *Sociology of Family Life*. Basingstoke: Palgrave Macmillan.

Clark, D. and Morgan, D. (1992). The Gaze of the Counsellors. S. Scott et al. (eds) *Public Risks and Private Dangers*. Aldershot: Avebury.

Collier, R. and Sheldon, S. (2008). *Fragmented Fatherhood: A Socio-Legal Study*. Oxford and Portland: Oregon Hart Publishing.

Connell, R. W. (1987). *Gender and Power*. Cambridge: Polity.

Craib, I. (1976). *Existentialism and Sociology: A Study of Jean-Paul Sartre*. Cambridge: Cambridge University Press.

Craib, I. (1995). Some Comments on the Sociology of Emotions. *Sociology*, 29(1) 151–58.

CRFR (Centre for Research on Families and Relationships) (2004). *Caringscapes: Experience of Caring and Working*. CRFR Research Briefing 13: University of Edinburgh.

Crompton, R. (2006). *Employment and the Family: The Reconfiguration of Work and Family Life in Contemporary Societies*. Cambridge: Cambridge University Press.

Cliff, D. (1993). Under the Wife's Feet: Renegotiating Gender Divisions in Early Retirement. *Sociological Review*, 41(1) 30–53.

Cunningham-Burley, S, Backett-Milburn, K, and Kemmer, D. (2005). Balancing Work and Family Life: Mother's Views. In L. McKie and S. Cunningham-Burley (eds) *Families in Society: Boundaries and Relationships*. Bristol: The Policy Press.

Curtis, P, James, A. and Ellis, K. (2009). 'She's Got a Really Good Attitude to Healthy food … Nannan's Drilled it into Her'. Intergenerational Relations within Families. In P. Jackson (eds) *Changing Families, Changing Food*. Basingstoke: Palgrave Macmillan, 77–92.

Dallos, S. and Dallos, R. (1997). *Couples, Sex and Power: The Politics of Desire*. Buckingham: Open University Press.

Daly, K. (1996). *Families and Time: Keeping Pace in a Hurried Culture*. London: Sage.

Davidoff, L. and Hall, C. (1987). *Family Fortunes*. London: Hutchinson.

Davies, M. (2010). *Moving Images: The Practices and Politics of Displaying Family Photographs*. Ph.D. Thesis: Keele University.

DeNora, T. (2000). *Music in Everyday Life*. Cambridge: Cambridge University Press.

Dermott, E. (2008). *Intimate Fatherhood: A Sociological Analysis*. London: Routledge.

DeVault, M. L. (1991). *Feeding the Family: The Social Organization of Caring as Gendered Work*. Chicago; University of Chicago Press.

Devine, F. (2004). *Class Practices: How Parents Help Their Children Get Good Jobs.* Cambridge: Cambridge University Press.

Dex, S. (2004). Work and Families. In J. Scott, J. Treas and M. Richards (eds) *The Blackwell Companion to the Sociology of Families,* Oxford: Blackwell, 435–56.

Dubin, R. (1962). Industrial Workers' Worlds: A Study of the 'Central Life Interests' of Industrial Workers. In A. M. Rose (ed.) *Human Behaviour and Social Processes: An Interactional Approach,* London: Routledge and Kegan Paul, 247–66.

Duncan, S. and Edwards, R. (1999). *Lone Mothers, Paid Work and Gendered Moral Rationalities.* Basingstoke: Palgrave Macmillan.

Duncombe, J. and Marsden, D. (1998). 'Stepford Wives' and 'Hollow Men'? Doing Emotion Work, doing Gender and 'Authenticity' in intimate Heterosexual Relationships. In G. Bendelow and S. J. Williams (eds), *Emotions in Social Life: Critical Themes and Contemporary Issues,* London: Routledge, 211–27.

Durkheim, E. (1965[1917]). *Suicide: A Study in Sociology.* London: Routledge and Kegan Paul.

Emslie, C. and Hunt, K. (2009). 'Live to Work' or 'Work to Live'? A Qualitative Study of Gender and Work-Life Balance among Men and Women in Mid-Life. *Gender, Work and Organization* 16(1) 151–72.

Erel, U. (2002). Reconceptualizing Motherhood: Experiences of Migrant Women from Turkey Living in Germany. In D. Bryceson and U. Vuorela (eds) *The Transnational Family: New European Frontiers and Global Networks,* Oxford: Berg, 127–46.

Esping-Andersen, G. (1999). *Social Foundations of Post-Industrial Economics.* Oxford: Oxford University Press.

Finch, J. (1989). *Family Obligations and Social Change.* Cambridge: Polity.

Finch, J. (2007). Displaying Families. *Sociology,* 41(1) 65–81.

Finch, J. (2008). Naming Names: Kinship, Individuality and Personal Names. *Sociology,* 42(4) 709–25.

Finch, J. and Mason, J. (1993). *Negotiating Family Responsibilities.* London: Routledge.

Finch, J. and Mason, J. (2000). *Passing On: Kinship and Inheritance in England* London: Routledge.

Ford, F. and Fraser, R. (2009). Off to a Healthy Start: Food Support Benefits of Low-income Women in Pregnancy. In P. Jackson (ed.) *Changing Families, Changing Food,* Basingstoke: Palgrave Macmillan, 19–34.

Fowler, B. Pierre Bourdieu. In A. Elliott and B. S. Turner *Profiles in Contemporary Social Theory,* London: Sage, 315–26.

Gabb, J. (2008). *Researching Intimacy in Families.* Basingstoke: Palgrave Macmillan.

Gambles, R., Lewis, S. and Rapaport, R. (2006). *The Myth of Work-Life Balance: The Challenge of Our Time for Men, Women and Societies.* Chichester: J. Wiley.

Garfinkel, H. (1967). *Studies in Ethnomethodology.* New Jersey: Prentice-Hall.

Gatrell, C. (2005). *Hard Labour: The Sociology of Parenthood.* Maidenhead: Open University Press.

Giddens, A. (1984). *The Constitution of Society: Outline of the theory of Structuration.* Cambridge: Polity Press.

Giddens, A. (1992). *The Transformation of Intimacy: Sexuality, Love and Eroticism in Modern Societies.* Cambridge: Polity Press.

Gillis, J. (1996). *A World of Their Own Making: Myth, Ritual and the Quest for Family Values.* Cambridge: Harvard University Press.

Glucksmann, M. (1990). *Women Assemble: Women Workers and the New Industries in Inter-War Britain*. London: Routledge.

Glucksmann, M. A. (1995). Why 'Work'? Gender and the 'Total Social Organization of Labour'. *Gender, Work and Organization*, 2(2) 63–75.

Glucksmann, M. (2000). *Cottons and Casuals: The Gendered Organization of Labour in Time and Space*. Durham: Sociologypress.

Glucksmann, M. (2005). Shifting Boundaries and Interconnections: Extending the 'Total Social Organization of Labour'. In L. Pettinger, J. Parry, R. Taylor and M. Glucksmann (eds) *A New Sociology of Work?* Oxford: Blackwell 19–36.

Goffman, E. (1967). *Interaction Ritual: Essays on Face-to-Face Behaviour*. New York: Doubleday Anchor.

Goffman, E. (1971). *Relations in Public*. London: Allen Lane, The Penguin Press.

Goode, W. J. (1971). Force and Violence in the Family. *Journal of Marriage and the Family*, 33(4) 624–36.

Gregory, A. and Milner, S. (2009). Editorial: Work-Life Balance: A Matter of Choice? *Gender, Work and Organization*, 16(1) 1–13.

Guillaume, C. and Pochic, S. (2009). What Would You Sacrifice? Access to Top Management and Work-Life Balance. *Gender, Work and Organization*, 16(1) 14–36.

Hakim, C. (2000). *Work-Lifestyle: Choices in the Twenty-First Century: Preference Theory*. Oxford: Oxford University Press.

Hakim, C. (2005). Sex Differences in Work-Life Balance Goals. In D. M. Houston (ed.) *Work-Life Balance in the Twenty-First Century*, Basingstoke: Palgrave Macmillan, 55–79.

Haldrup, M. and Larsen, J. (2003). The Family Gaze, *Tourist Studies* 3(1) 23–46.

Halrynjo, S. (2009). Men's Work-Life Conflict: Career, Care and Self-Realization, Patterns of Privileges and Dilemmas. *Gender, Work and Organization* 16(1) 98–125.

Haugen, G. M. D. (2007). *Divorce and Post-Divorce Family Practices: The Perspectives of Children and Young People*, Trondheim Norwegian Centre for Child Research, NTNU, doctoral Thesis.

Hekman, S. J. (1995*). Moral Voices, Moral Selves: Carol Gilligan and Feminist Moral Theory*. Cambridge: Polity.

Hobson, B. and Morgan, D. (2002). Introduction: Making Men into Fathers. In B. Hobson (ed.) *Making Men into Fathers: Men, Masculinities and the Social Politics of Fatherhood*, Cambridge: Cambridge University Press 1–24.

Hochschild, A. R. (1983). *The Managed Heart: Commercialization of Human Feeling*. Berkeley: University of California Press.

Hochschild, A. R. (1998). The Sociology of Emotion as a Way of Seeing. In G. Bendelow and S. J. Williams (eds) *Emotions in Social Life: Critical Themes and Contemporary Issues*, London: Routledge 3–15.

Hochschild, A. R. (2003). *The Commercialization of Intimate Life: Notes from Home and Work*. Berkeley: University of California Press.

Hockey, J., Meah, A. and Robinson, V. (2007). *Mundane Heterosexualities: From Theory to Practices*. Basingstoke: Palgrave Macmillan.

Hogarth, T. and Daniel, W. W. (1988). *Britain's New Industrial Gypsies: Long Distance Weekly Commuters*. London: Policy Studies Institute.

Holdsworth, C. and Morgan, D. (2005). *Transitions in Context: Leaving Home, Independence and Adulthood*. Maidenhead: Open University Press.

Holdsworth, C. and Morgan, D. (2007). Revisiting the Generalized Other: An Exploration. *Sociology* 41(3) 401–17.

Houston, D. M. (2005). *Work-Life Balance in the Twenty-First Century*. Basingstoke: Palgrave Macmillan.

Hughes, E. (1971). *The Sociological Eye*. Chicago: Aldine Atherton.

Hyman, J., Scholarios, D. and Baldry, C. (2005). 'Daddy, I Don't Like Those Shifts You're Working Because I Never See You': Coping Strategies for Home and Work. In D. M. Houston (ed.) *Work-Life Balance in the Twenty-First Century* Basingstoke: Palgrave Macmillan, 122–46.

Isaksen, L. W. (2005). Gender and Care: The Role of Cultural Ideas of Dirt and Disgust. In D. Morgan, B. Brandth and E. Kvande (eds) *Gender, Bodies and Work*, Aldershot: Ashgate 115–26.

Jackson, P. (2009). *Changing Families, Changing Food*. Basingstoke: Palgrave Macmillan.

James, A. (2000). Embodied Being(s): Understanding the Self and the Body in Childhood. In A. Prout (ed.) *The Body, Childhood and Society*, Basingstoke: Palgrave Macmillan 19–37.

Jamieson, L. (1998). *Intimacy: Personal Relationships in Modern Societies*. Cambridge: Polity.

Jordan, B., Redley, M. and James, S. (1994). *Putting the Family First: Identities, Decisions, Citizenship*. London: UCL Press.

Kvande, E. (2005). Embodying Male Workers as Fathers in a Flexible Working Life. In D. Morgan, B. Brandth and E. Kvande (eds) *Gender, Bodies and Work*, Aldershot: Ashgate 75–88.

Kvande, E. (2007). *Doing Gender in Flexible Organizations*. Bergen: Fagbokforlaget.

Lammi-Taskula, J. (2007). *Parental Leave for Fathers? Gendered Conceptions and Practices in Families with Young Children in Finland*. Finland STAKES Report 166.

Le Bihan, B. and Martin, C. (2008). Caring for Dependent Elderly Parents and Family Configurations. In E. D. Widmer and R. Jallinoja (eds) *Beyond the Nuclear Family: Families in a Configurational Perspective*, Bern: Peter Lang 59–76.

Lefebvre, H. (1991). *The Production of Space*. Oxford: Basil Blackwell.

Manning, I. (1978). *The Journey to Work*. Sydney: Allen and Unwin.

Mason, J. (1999). *Inheriting Money: Kinship and Practical Ethics*. University of Leeds, CRFKC Working Paper 13.

Mason, J. (2004). Managing Kinship over Long Distances: The Significance of the Visit. *Social Policy and Society* 3(4) 421–9.

Mason, J., May, V. and Clarke, L (2007). Ambivalence and the Paradoxes of Grandparenting. *Sociological Review* 55(4) 687–706.

Mayall, B. (1998). Children, Emotions and Daily Life at Home and School. In G. Bendelow and S. J. Williams (eds) *Emotions in Social Life: Cultural Themes and Contemporary Issues*. London: Routledge 135–54.

McKie, L., Bowlby, S. and Gregory, S. (2004). Starting Well: Gender, Care and Health in the Family Context. *Sociology* 38(3) 593–612.

McKie, L. and Cunningham-Burley, S. (2005). *Families in Society: Boundaries and Relationships*. Bristol: The Policy Press.

McKie, L., Gregory, S. and Bowlby, S. (2002). Shadow Times: The Temporal and spatial Frameworks and Experiences of Caring and Working. *Sociology* 36(4) 897–924.

Mead, G. H. (1962[1934]). *Mind, Self and Society*. Chicago: University of Chicago Press.

Meah, A., Hockey, J. and Robinson, V. (2008). What's Sex Got to Do with it? A Family-Based Investigation of Growing up Heterosexual During the Twentieth Century. *Sociological Review* 56(3) 454–73.

Metcalfe, A., Dryden, C., Johnson, M., Owen, J. and Shipton, G. (2009). Fathers, Food and Family Life. In P. Jackson (ed.) *Changing Families, Changing Food*, Basingstoke: Palgrave Macmillan 93–117.

Miller, T. (2010). *Making Sense of Fatherhood: Gender, Caring and Work*. Cambridge: Cambridge University Press.

Mills, C. W. (1959). *The Sociological Imagination*. Oxford: Oxford University Press.

Minkler, M. and Estes, C. L. (1991). *Critical Perspectives of Aging: The Political and Moral Economy of Growing Old*. New York: Baywood Publishing Co. Inc.

Morgan, D. H. J. (1975). *Social Theory and the Family*. London: Routledge and Kegan Paul.

Morgan, D. H. J. (1985). *The Family, Politics and Social Theory*. London: Routledge and Kegan Paul.

Morgan, D. H. J. (1996). *Family Connections: An Introduction to Family Studies*. Cambridge: Polity.

Morgan, D. (2001). Family Sociology in From the Fringe: The Three 'Economies' of Family Life. In R. G. Burgess and A. Murcott (eds) *Developments in Sociology*. Edinburgh: Pearson Education, 227–48.

Morgan, D. (2002). Sociological Perspectives on the Family. In A. Carling, S. Duncan and R. Edwards (eds) *Analysing Families: Morality and Rationality in Policy and Practice*, London: Routledge 147–64.

Morgan, D. (2004). Everyday Life and Family Practices. In E. B. Silva and T. Bennett (eds) *Contemporary Culture and Everyday Life*, Durham: Sociologypress 37–51.

Morgan, D. (2009). *Acquaintances: The Space between Intimates and Strangers*. Maidenhead: Open University Press.

Morgan, D., Brandth, B. and Kvande, E. (2005). *Gender, Bodies and Work*. Aldershot: Ashgate.

Nussbaum, M. C. (2001). *Upheavals of Thought: The Intelligence of Emotions*. Cambridge: Cambridge University Press.

Obama, B. (2007). *Dreams from My Father*. Edinburgh: Canongate.

Oinonen, E. (2008). *Families in Converging Europe: A Comparison of Forms, Structures and Ideals*. Basingstoke Palgrave Macmillan.

Pahl, R. (1984). *Divisions of Labour*. Oxford: Basil Blackwell.

Parry, J., Taylor, R., Pettinger, L. and Glucksmann, M. (2005). Confronting the Challenges of Work Today: New Horizons and Perspectives. In L. Pettinger, J. Parry, R. Taylor and M. Glucksmann (eds) *A New Sociology of Work?* Oxford: Blackwell 3–18.

Phillipson, C., Allan, G. and Morgan, D. (2004). *Social Networks and Social Exclusion: Sociological and Policy Perspectives*. Aldershot: Ashgate.

Pickvance, C. G. and Pickvance, K. (1995). The Role of Family Help in the Housing Decisions of Young People *Sociological Review* 43(1) 123–49.

Prout, A. (ed). (2000). *The Body, Childhood and Society*. Basingstoke: Macmillan.

Reay, D. (2005). Doing the Dirty Work of Social Class? Mother's Work in Support of Their Children's Schooling. In L. Pettinger, J. Parry, R. Taylor and M. Glucksmann (eds) *A New Sociology of Work?* Oxford: Basil Blackwell, 104–16.

Reckwitz, A. (2002). Toward a Theory of Social Practices: A Development in Culturalist Theorizing. *European Journal of Social Theory*, 5(2) 243–63.

Reiger, K. (1985). *The Disenchantment of the Home: Modernizing the Australian Family, 1880–1960*. Oxford: Oxford University Press.

Ribbens McCarthy, J., Edwards, R. and Gillies, V. (2003). *Making Families: Moral Tales of Parenting and Step-Parenting*. Durham: Sociologypress.

Richards, L. (1990). *Nobody's Home: Dreams and Realities in a New Suburb*. Oxford: Oxford University Press.

Roberts, B. (2002). *Biographical Research*. Buckingham: Open University Press.

Roberts, E. (2008). Time and Work-Life Balance: The Roles of 'Temporal Customization' and 'Life Temporality'. *Gender, Work and Organization*, 15(5) 430–53.

Roper, M. (1994). *Masculinity and the British Organization Man Since 1945*. Oxford: Oxford University Press.

Rosen, M/Bloomfield, J. (1997). All in the Family. In P. Cohen (ed.) *Children of the Revolution: Communist Childhood in Cold War Britain*. London: Lawrence and Wishart, 52–77.

Roseneil, S. (2005). Living and Loving Beyond the Boundaries of the Heteronorm: Personal Relationships in the Twenty-First Century. In L. McKie and S. Cunningham-Burley(eds) *Families in Society: Boundaries and Relationships*. Bristol: The Policy Press 241–58.

Savage, M., Barlow, J., Dickens, P. and Fielding, T. (1992). *Prosperity, Bureaucracy and Culture: Middle-Class Formation in Contemporary Britain*. London: Routledge.

Sayer, A. (2005). *The Moral Significance of Class*. Cambridge: Cambridge University Press.

Schatzki, T. R. (1996). *Social Practices: A Wittgenstinnian Approach to Human Activity and the Social*. Cambridge: Cambridge University Press.

Sevenhuijsen, S. (1998). *Citizenship and the Ethics of Care: Feminist Considerations on Justice, Morality and Politics*. London: Routledge.

Sevenhuijsen, S. (1999). *Caring in the Third Way*. Leeds: CRFKC Working Paper 12.

Seymour, J. (2007). Treating the Hotel Like a Home: The Contribution of Studying the Single Location Home/Workplace. *Sociology* 41(6) 1097–114.

Silva, E. B. (2004). Materials and Morals: Families and Technologies in Everyday Life. In E. B. Silva and T. Bennett (eds) *Contemporary Culture and Everyday Life*, Durham: Sociologypress 52–70.

Smart, C. (2007). *Personal Life*. Cambridge: Polity.

Smart, C. and Neale, B. (1999). *Family Fragments?* Cambridge: Polity.

Smith, A. (1759[1976]). *The Theory of Moral Sentiments*. Oxford: Clarenden Press.

Smith, D. E. (1987). *The Everyday World as Problematic: A Feminist Sociology*. Toronto: University of Toronto Press.

Smith, D. E. (1993). The Standard North American Family: SNAF as an Ideological Code. *Journal of Family Issues*, 14(3) 50–65.

Southerton, D. (2006). Analysing the Temporal Organization of Daily Life: Social Constraints, Practices and their Allocation. *Sociology*, 40(3) 435–54.

Spencer, L. and Pahl, R. (2006). *Rethinking Friendship: Hidden Solidarities Today*. Princeton: Princeton University Press.

Stanley, L. (1995). *The Auto/Biographical I*. Manchester: Manchester University Press.

Taylor, M., Houghton, S. and Durkin, K. (2008). Getting Children with Attention Deficit Hyperactivity Disorder to School on time: Mothers' Perspectives. *Journal of Family Issues* 29(7) 918–43.

Tober, D. M. (2001) Semen as Gift, Semen as Goods: Reproductive Workers and the Market in Altruism. *Body and Society* 7(2/3) 137–60.

Ungerson, C. and Yeandle, S. (2005). Care Workers and Work-Life Balance: The Example of Domiciliary Careworkers. In D. M. Houston (ed.) *Work-Life Balance in the Twenty-First Century*. Basingstoke: Palgrave Macmillan, 246–62.

Urry, J. (2002). Mobility and Proximity. *Sociology*, 36(2) 255–74.

Vincent, C. and Ball, S. J. (2006). *Children, Choice and Class Practices*. London: Routledge.

Wall, K., Aboim, S. and Marinho, S. (2008). Family Configurations from the Male Perspective: Exploring Diversity Over the Life Course. In E. D. Widmer and R. Jallinoja (eds) *Beyond the Nuclear Family: Families in a Configurational Perspective*. Bern: Peter Lang 207–29.

Weeks, J., Heaphy, B. and Donovan, C. (2001). *Same Sex Intimacies: Families of Choice and Other Life Experiments*. London: Routledge.

Weigt, J. M. and Soloman, C. R. (2008). Work-Family Management Among Low-Wage Service Workers and Assistant Professors in the USA: A Comparative Institutional Analysis. *Gender, Work and Organization* 15(6), 621–49.

Widerberg, K. (2005). Embodied Gender Talks – The Gendered Discourse of Tiredness. In D. Morgan, B. Brandth and E. Kvande (eds) *Gender, Bodies and Work*, Aldershot: Ashgate, 101–13.

Widmer, E. D. et al. (2008). Introduction. In E. D. Widmer and R. Jallinoja (eds) *Beyond the Nuclear Family: Families in a Configurational Perspective*, Bern: Peter Lang, 1–11.

Widmer, E. D. and Jallinoja, R. (2008). *Beyond the Nuclear Family: Families in a Configurational Perspective*. Bern: Peter Lang.

Widmer, E. D. and Sapin, M. (2008). Families on the Move: Insights on Family Configurations of Individuals Undergoing Psychotherapy. In E. D. Widmer and R. Jallinoja (eds) *Beyond the Nuclear Family: Families in a Configurational Perspective*, Bern: Peter Lang 279–302.

Wolkowitz, C. (2009). Challenging Boundaries: An Autobiographical Perspective on the Sociology of Work. *Sociology*, 43(5) 846–60.

Index